APPLICATIONS FOR ELECTRONIC DISPLAYS

APPLICATIONS FOR ELECTRONIC DISPLAYS

Technologies and Requirements

SOL SHERR

A Wiley-Interscience Publication

JOHN WILEY & SONS, INC.

New York • Chichester • Weinheim • Brisbane • Singapore • Toronto

Copyright ©1998 by John Wiley & Sons, Inc. All rights reserved.

Published simultaneously in Canada.

Library of Congress Cataloging in Publication Data:

Sherr, Sol
    Applications for electronic displays : technologies and
requirements / Sol Sherr.
        p.   cm.
    Includes index.
    ISBN 0-471-04228-5 (cloth : alk. paper)
    1. Information display systems.   I. Title
TK7882.I6S485   1998
621.3815′422—dc21   97-26261

Printed in the United States of America

10 9 8 7 6 5 4 3 2 1

CONTENTS

PREFACE   vii

PART I   TECHNOLOGIES AND REQUIREMENTS   1

1   Applictions Review and Human Visual Interface
     Requirements   3
     *Major electronic display applications, visual requirements,
     visual and photometric parameters, and a representative
     display specification.*

2   Human Factors Considerations and Technology Review   22
     *Human factor considerations, photometric and visual
     parameters, display technologies, CRT design electronics,
     FPD design considerations, and matrix and multiplexing
     drive techniques.*

PART II   PRODUCT APPLICATIONS   85

3   Output Devices and Systems   87
     *Major output devices and systems, parameter tables for CRT
     and FPD monitors, FPD panels, hard-copy devices, and
     large-screen systems.*

4   Systems and User Applications   166
     *Computer terminals, workstations, host–terminal systems,
     display application specifications, and display technology
     evaluation for applications.*

PART III   **USER APPLICATIONS**   **205**

5   **Computer Graphics**   **207**
Computer graphics applications, display specification requirements for applications and available display technologies that best meet display specification requirements.

6   **Visualization and Imaging (VISIM)**   **229**
Display requirements and specifications for CAD/CAE, documents, earth resources, geographic information systems, mathematics, medical, presentations, scientific data, and user-driven technology parameters.

7   **Multimedia and Presentations**   **255**
Special display requirements, specifications, and technology performance comparison charts for multimedia and presentation applications.

8   **Entertainment: Electronic Games Television, and Video**   **274**
Display requirements for electronic games, television, video systems, technology evaluation and comparison charts, and system block diagrams

9   **Computer Aided Design, Engineering, and Manufacturing (CAD/CAE/CAM)**   **289**
Applications descriptions, display performance requirements, operating characteristics, analysis of technologies, graphics workstation, and system block diagrams.

10.   **Virtual Reality**   **313**
Virtual reality applications, systems block diagrams, software, equipment, HUD, HMD requirements, table of applications and requirements, parameters and technologies.

11   **Summary**   **335**
List of applications and display technologies, best product types, specifications, matrix charts, and numeric ratings by technologies and product types.

INDEX   **347**

PREFACE

This book is a companion volume and an expansion of the second edition of *Electronic Displays*. As noted in the Preface to *Electronic Displays*, 2nd ed. the chapter on applications in *Electronic Displays*, 1st ed. was replaced in the second edition by a chapter on input and output devices, with the comment that the subject of electronic display applications might warrant a separate book in the not-too-distant future. This volume is that book, although it took somewhat longer than expected to become a reality. The delay has led to a more complete coverage of the subject, however, as electronic display applications have expanded to such an extent that they now represent a highly significant aspect of computer usage such as computer-aided design, engineering, and manufacturing (CAD/CAE/CAM), computer graphics, entertainment (most commonly television and video games), multimedia, presentations, and virtual reality, to name some of the most important examples. It should be noted, however, that although this text has been planned to expand on the more limited treatment of applications found in *Electronic Displays*, 1st. ed., and can be used effectively in conjunction with *Electronic Displays*, 2nd ed., it is also intended as a standalone volume, to be consulted as an independent source of information on the display technologies used to produce the electronic displays involved in the applications covered in the book. To this end, several chapters are devoted to discussion and exposition of these technologies, and reading of the earlier text is not required to obtain an adequate understanding of those technologies. Of course, for a thorough understanding of the subject, the reader is advised to refer to both editions of *Electronic Displays*, but those readers who are primarily interested in the applications per se. may skip these earlier volumes without undue lack of information.

In order to achieve these objectives, the book is divided into 11 chapters, each containing illustrations that clarify the text, as well as references that provide other sources of information on the subjects covered in each chapter. The chapters are further grouped into three main parts entitled "Applications Review, Technologies, and Visual Requirements," "Product

Applications," and "Specific User Applications," respectively. These main sections in turn are divided into the 11 chapters as follows: applications review and human interface requirements (Chapter 1), human factors considerations and technology review (Chapter 2), in Part 1: output devices and systems (Chapter 3), and data processing systems, input devices, and workstations (Chapter 4); in Part 2; computer graphics (Chapter 5), visualization and imaging (Chapter 6), multimedia and presentations (Chapter 7), entertainment (Chapter 8), computer aided design, engineering, and manufacturing (Chapter 9), and virtual reality (Chapter 10), in Part 3; Chapter 11 summarizes, recapitulates, and reviews the material presented in the previous chapters. Some of the material presented in *Electronic Displays*, 2nd ed. is repeated in Chapters 2 and 3 of the present volume, but with an application orientation point of view. The rest of the text is completely new, and as up-to-date in its content and subject matter as possible. Thus, I am confident that this volume will be of significant value to all engineers and related professionals who are involved to any significant extent in the field of electronic display technologies and applications. It is my sincere hope that those individuals who obtain and use this text will be gratified by the results, and that a second edition will be found necessary sometime before the budget is balanced.

Before ending this preface, I should like to express my appreciation to several individuals whose encouragement and support have been so important during those long, lonely days of diligent effort. First and foremost, I must pay proper respects to my wife, Claire, whose understanding and acceptance of the time consumed by this effort have been a source of strength. Second, I should like to express my thanks to my good friend and coworker in the field, that well-known international expert on computer graphics, Carl Machover, whose suggestions and information have been of special value. Finally, I should like to thank George Telecki, my editor at John Wiley, for his unfailing good humor during some rather difficult times, and constant assistance in overcoming the difficulties; Mary Lynn, also from Wiley, for her invaluable help in obtaining permission to use the many illustrations from a variety of sources; and Rosalyn Farkas for the outstanding job of guiding the manuscript through production. The work couldn't have been done so expeditiously without their help.

SOL SHERR

Old Chatham, NY

APPLICATIONS FOR ELECTRONIC DISPLAYS

PART I

TECHNOLOGIES AND REQUIREMENTS

1

APPLICATIONS REVIEW
AND HUMAN VISUAL
INTERFACE REQUIREMENTS

1.1 INTRODUCTION

There are numerous user-oriented applications for electronic displays, ranging from the simplest readouts to the most complex computer-aided design/manufacturing (CAD/CAM) and virtual reality displays. Each of these applications has its own requirement for the characteristics of the visual interface; these characteristics, in turn, affect the requirements for the technology that is used to provide the display. This chapter briefly describes and reviews all of the most and some of the less significant applications, to be elaborated on in later chapters. The requirements imposed by the user's visual and perceptual limitations, commonly referred to as "human factors" are discussed next. The human factors discussion begins with a list and short definitions of relevant visual parameters in terms of the capabilities of the human visual system. This is followed by a review of the photometric parameters that are used to describe the various aspects of visual display systems, as measured by appropriate instruments, and brief descriptions of how the instruments are used. The goal of these discussions is to provide the reader with adequate understanding of how visual performance is specified and measured so that the manufacturer's specifications can be properly interpreted.

The section on human factors ends with the presentation of a representative specification for the "fine art" application, with a description of how this application uses the display portion of the applicable equipment and systems. This short, preliminary discussion is supplemented by the more detailed material found in the relevant chapters, and should be quite adequate as an introduction. Many of the applications do not require detailed discussion at this point and are covered in depth in later chapters. The treatment at this

3

point should suffice to allow the reader to understand the relationship between display applications and visual and performance specifications for the display devices and systems used for particular applications. It is important to recognize that the display visual characteristics facilitate the effective utilization or the enjoyment—in the entertainment and virtual reality applications—of the information presented visually. With this brief introduction, it should be possible for the reader to understand the relationship between a specific application and the type of display device and technology used for that application.

1.2   APPLICATIONS

All the major and some of the minor applications that use one or more types of visual display as the main means of interaction between the human user and the operating system are shown in Table 1.1, which lists the requirements and types limited to a few generalized groupings; these applications are presented in greater detail in later chapters.

These applications make up a fairly complete list of those that use visual displays in some form to carry out the operation and establish a basis for determining what the requirements and specifications should be for the display equipment, system, and devices. Thus, the next task is to define the parameters involved and describe the application so that a set of values and a procedure for selecting these values may be derived for each application involved. This is done in a preliminary manner in following sections, and in greater detail in subsequent chapters.

1.3   APPLICATION DESCRIPTIONS

1.3.1   Introduction

The descriptions contained in this section are brief and introductory; more detailed descriptions are found in later chapters devoted to specific individual and groups of applications. The purpose of this chapter is to prepare for the development of preliminary display requirements and specifications for the various applications, present one specification to illustrate the process, and discuss the display technologies in terms of how well each one can meet the particular applications. This is done in detail in Chapter 2, but some preliminary material is contained here as well. Not all the applications are of equal importance, and the length of each description is related to the relative importance of the application. For example, CAD is a very important application and is allotted one of the greater amounts of space, whereas art is still very limited in use. Conversely, measurements is a widespread applica-

TABLE 1.1   List of Display Applications

Application	Display Requirements	Display Type
Advertising	Graphics, color, high resolution	Large-board, projection
Animation	Graphics, color, high resolution	CRT, flat-panel
Computer-aided design (CAD)	Graphics, color, high resolution	CRT, flat-panel, plotter
Computer-aided engineering (CAE)	Graphics, high resolution	CRT, flat-panel
Computer-aided manufacturing (CAM)	Graphics, medium resolution	CRT, flat-panel
Computer-integrated manufacturing (CIM)	Graphics, color, low resolution	CRT, flat-panel, readout
Computer graphics	Graphics, color, high resolution	CRT, flat-panel, projection
Communication	Alphanumeric capability (A/N)	Flat-panel, readout
Control	A/N, graphics, color	CRT, flat-panel, readouts
Data processing	Numeric	CRT, flat-panel, printer
Desktop publishing	A/N, graphics, high resolution	CRT, Flat-panel, printer
Document preparation	A/N, graphics	CRT, flat-panel, printer
Education	A/N, Graphics, color high reolution	CRT, flat-panel, projection
Electronic games	Graphics, color, high resolution	CRT, flat-panel
Entertainment	Graphics, color, high resolution	CRT, flat-panel, projection
Geographic systems	A/N, graphics, color, high resolution	CRT, flat-panel readout
Imaging systems	A/N, graphics, color high resolution	CRT, flat-panel, projection
Manufacturing	A/N, graphics, color, low resolution	CRT, flat-panel, readout
Measurements	A/N, color, medium resolution	CRT, flat-panel, plotter
Message displays	A/N, graphics, medium resolution	CRT, projection, large-board
Military	A/N, graphics, color, high resolution	CRT, flat-panel projection
Monitoring	A/N, graphics	CRT, flat-panel readout
Multimedia	Graphics, color, high resolution	CRT, flat-panel
Navigation	A/N, graphics, color	CRT, flat-panel, readout
Presentations	A/N, graphics, color, high resolution	CRT, flat-panel, projection

TABLE 1.1 (*Continued*)

Application	Display Requirements	Display Type
Simulation	Graphics, color, high resolution	CRT, flat-panel, projection
Sports	A/N, graphics, color, high resolution	Flat-panel, projection
Television	Graphics, color, high resolution	CRT, flat-panel, projection
Transportion	A/N, graphics, color, high resolution	CRT, flat-panel, large-board
Utilities	A/N, graphics, color, high resolution	CRT, flat-panel, large-board
Virtual reality	Graphics, color, high resolution	CRT, flat panel

tion, but the displays tend to be rather simple, usually no more than an A/N unit. Similar considerations apply to the other applications covered in the next section.

1.3.2 Descriptions

1.3.2.1 Advertising

The use of electronic displays in advertising is rather limited, but not wholly insignificant. One of the main features is the inclusion of animation in some of the most effective outdoor billboards. These billboards are surprisingly effective when one considers the limited resolution that can be achieved with the usual matrix of standard lightbulbs that can be assembled. This resolution seldom exceeds 100×100 because of the large power requirements of standard light bulbs. However, some new versions of this type of board use small CRTs assemblies; a much larger number of these displays can be assembled as they are not subject to the power limit that afflicts lightbulbs. Examples of this type of large-board displays are those produced by Philips, Sony, and others and designated videowall. These displays use various approaches, such as assemblies of individual CRTs or projectors, and are discussed in detail later. These units can display TV-quality video, and make ideal large-board displays for advertising purposes. Another type of nonprojection large board that can be used inside or outside for advertising is that which uses large matrix assemblies of LEDs (light emitting diodes). These function very much like the standard lightbulb varieties, but differ in that the power requirement is much lower and therefore much larger assemblies are feasible. In summary, much of the advertising that uses some form of electronic display is found in outdoor billboards or animated signs. The requirements stress high visibility in difficult environments; thus the display technology must be capable of high light output. The relatively low resolution and light output capabilities found in earlier models have been replaced with new technologies that have made higher resolution and light output achievable. The indoor displays may have only moderate requirements in these parameters.

1.3.2.2 Animation

Computer-generated animation has superseded the manual techniques in recent years. As a result, the specifications for the display device parameters have become increasingly complex in computer display systems used for animation. In particular, the requirements for resolution and the number of colors have become quite stringent, especially when the result is to be an animated cartoon for cinematic presentation. This contrasts with the much cruder animation that is used in advertising, as noted previously. However, the major animation activities are for movies and videos, so that the more rigorous requirements are the significant ones, and the technologies employed must be those that allow the display system to meet these requirements. In general, CRTs are most effective, but LCDs (liquid crystal displays) are beginning to meet these requirements, with plasma, and several new technologies also showing some promise.

1.3.2.3 Art

The use of computer-generated displays for the production of fine art has been carried out with varying degrees of success for a number of years. Initially, the major output device was a plotter with varying degrees of capability in terms of resolution and color. Plotters have reached a level of performance that is adequate for many of the requirements of this type of art, and the artistic use of computer-generated images has been extended to include video representations other than the animated cartoons discussed previously. Thus, there is an expanded market for computer-generated art, and this is an applications that, while not large, is still significant. As such, it should be considered as having some impact on what the capabilities of display technologies used for these purposes need to be. For this particular application, the combination of CRTs and high-performance plotters appears to be most successful.

1.3.2.4 Avionics

Avionics is a catchall application category that applies to any electronically generated display that is used for some type of airborne application. The main application is for the cockpit display used for navigation purposes and is usually located directly in front of the pilot, but head-up, head-down, and helmet-mounted displays are also included. Other avionics applications are those involved in operating the NASA Space Shuttle, and these may include many special features. In general, the requirements for all of these displays range from simple A/N units to quite complex presentations, and the technologies similarly go from simple readouts to high-definition and multicolor CRT and flat-panel units. Thus, avionics is one of the most demanding of the display applications.

1.3.2.5 Business Systems

Business systems include all forms of business operations that are computer-based and function with some type of output devices to determine the results of the business-oriented operations and make them available for control and modification. These include various payroll and personnel functions, as well as training, presentations, and financial operations. As a result, a large number of individual displays may be required to cover all of these functions in a single business or plant.

1.3.2.6 CAD

Computer-aided design (CAD) is another of the most demanding applications for computer-generated electronic displays. Initially, it accounted for the bulk of the computer graphics application, and the electronic display requirements were among the most stringent. These requirements have not diminished over the years, but there are now numerous other applications that call for equal or greater requirements in terms of resolution, color, and quality in general. Thus, the most popular technology for systems designed to meet CAD requirements tend to emphasize the CRT, which is still the best display output device for attaining the highest visual quality. However, flat-panel displays are beginning to compete with CRTs, in particular the active- and passive-matrix LCDs, so perhaps flat-panel displays (FPDs) may be used for this application in the not-too-distant future.

1.3.2.7 CAE

Computer-aided engineering (CAE) is probably the oldest example of the use of computers to aid in the development and design of electrical and mechanical products. Initially, the display portion was restricted to the generation of graphs and tables on plotters, but present applications go far beyond this limited form of presentation of information in visual form, and use the full capabilities of computer-generated display systems. Examples of outputs include schematics, wiring diagrams, mechanical drawings, and other engineering-related displays. These and other similar displays generally require systems having the full range of resolution, color, and image quality; thus, CRTs are preferred, but flat-panel displays are beginning to meet those needs.

1.3.2.8 CAM

Computer-aided manufacturing (CAM) is a companion to CAD and is usually presented as part of the CAD/CAM package as though the two were inseparable. They are actually independent, insofar as the applications are concerned, with CAM the less demanding of the two. However, this does not mean that the CAM requirements for electronic display performance are minimal, and simple A/N units are by no means adequate. Indeed, the trend is toward more elaborate displays as the automatic aspects of manufacturing become more prevalent and introduce the need for more information on the factory floor in order to adequately monitor performance. Therefore, CRTs are the preferred choice, although flat-panels can meet many of the requirements.

1.3.2.9   CIM

Computer-integrated manufacturing (CIM) is the third component of CAD/CAM/CIM, but is more recent than the other two and far less common to date. As the term implies, this application uses computers to fully integrate the manufacturing process, avoiding the need for manual operations. Therefore, the displayed information is purely for monitoring purposes, and may be quite limited.

1.3.2.10   Computer Graphics

Computer graphics is the generic term that includes all the applications that call for highly complex visual presentations of a large variety of images, including all those used for CAD/CAE/CAM/CIM described previously. However, it does go well beyond these four areas and applies to all applications that use computers to generate graphic images. Therefore, it is possible to limit the discussion of display applications to computer graphics alone. In the interest of completeness, however, all subcategories of computer graphics are treated separately. The display requirements for the generic computer graphics application are the most stringent, and, if met, should be adequate for all subcategories. Those requirements include the best resolution, color gamut, and image quality obtainable with available technology. CRTs at present are needed to meet the most extreme requirements, with flat-panel units adequate for many lesser requirements.

1.3.2.11   Communication

This application includes all means for transmitting messages from one location to another, such as telephone, telegram, electronic mail, (e-mail), facsimile (fax), and voicemail. The majority of these have very simple display requirements, usually limited to A/N modules. However, with the growing use of video displays in communication systems, the display requirements are increasing and may meet those found for computer graphics. At present, only A/N modules are used for the majority of systems, consisting primarily of FPDs, but for video telephones the flat-panel display may not be adequate. In any event, the development of more complex systems in the future is surely anticipated.

1.3.2.12   Control

This is a wide-ranging application, including simple systems used for some manufacturing operations and the complex ones found in utility control and monitoring installations. The former requires little more than limited A/N modules, but the latter may be a large system with numerous CRT, flat-panel, and large-board displays. The requirements can be of the most advanced in each case, so that this application can involve extremely stringent visual specifications. All the technologies can be used to meet these extreme requirements, with the flat-panel technologies widely used for readouts and similar purposes, while the CRT is more prevalent for other information displays. However, FPDs are advancing rapidly in their capabilities and may supplant CRTs very soon as the preferred technologies for all control applications.

1.3.2.13   Data Processing

This application is essentially concerned with operating with numerics, and to some extent with alphas. Therefore, the main electronic display requirement is for A/N modules. In some cases it is desired to add graphics and other more complex visual images to the readouts, however, so there may be some need for more complex visual equipment. To meet this need, it may be necessary to add CRT or flat-panel units with at least a moderate capability.

1.3.2.14   Desktop Publishing

This is a broad classification that may include a wide variety of actual applications, such as books, brochures, and magazines. As a result, the actual requirements are quite varied, depending on the quality and type of results expected. For example, in the case of a simple in-house publications, the system may be relatively minimal and the display results equivalent, whereas when the result is to be a finished design and layout for a book to be published, the display requirements are correspondingly severe. This type of publishing application is taking over many of the publishing operations, and the total potential of the application is large.

1.3.2.15   Document Preparation

The display requirements for this application are largely those for any word processing application, but with the addition of the potential need for a variety of graphical illustrations such as charts and graphs. These requirements may be quite complex when high-quality brochures and reports are the main outputs. CRTs and flat-panels are both possible technologies for the preparation of these documents, supported by printers and plotters.

1.3.2.16   Education

Education is an application that is difficult to classify in terms of its electron display requirements, as it may be as simple as a small A/N numeric module, or as complex as a high-resolution, multicolor display. For the latter, CRTs are still best, as good color FPDs are still quite expensive when compared with high-quality CRT units. However, when the flat-panels begin to approach CRTs in terms of cost effectiveness as well as quality, they may become the preferred technology.

1.3.2.17 Electronic Games

Electronic games are particularly intriguing in that they may use very simple or highly complex graphics. The inexpensive, battery-operated types all use flat-panel displays based on LCDs, and provide rather minimal graphics. However, the very high-quality units that have complex graphics capabilities are still based primarily on CRTs. As FPDs become cheaper and more capable of high-quality color graphics, they should replace CRTs because of the advantages in portability and power. The requirements for the high-performance units will remain for maximum resolution and color capabilities.

1.3.2.18 Entertainment

Electronic games, television, video, multimedia, and virtual reality all might be listed under the *entertainment* category (which has a general connotation), but their unique characteristics and their occasional use in nonentertainment applications warrant their separate listing. The specific entertainment aspects are primarily those found in arcades and at home and are intended mainly for individual and group delectation, although they may also have some educational value. This is particularly true of television as a generalized application, and also virtual reality as a means for presenting visual images to highlight specific experiences. Given these differences, it seems appropriate to keep entertainment as a separate application, although its display requirements are basically those that apply to most of the other specific applications cited above. Therefore, it is not necessary to repeat the statements about the technologies involved in meeting these applications' display requirements.

1.3.2.19 Geographic Systems

Cartography, or mapmaking, is the most common activity for this application, but *geographical information systems* (GIS) is the more inclusive designation. However, it can be considered as referring to the generation of all types of map like images, and is not necessarily restricted to geographic ones in spite of the general designation. These maps can be quite complex, with multiple colors, although it has been demonstrated that for most geographic maps four or five colors will suffice. However, the resolution requirements can be very high and CRTs are the preferred technology. However, as in many other applications with similar requirements, FPDs are approaching the capabilities required at least for less than the most complicated maps.

1.3.2.20 Imaging Systems

Imaging systems represent a rather broad category containing a wide range of sources of fixed images that need to be presented in electronic form. Examples are various types of medical images such as x-rays and the data generated by computered axial tomography (CAT) and magnetic resonance imaging (MRI) scans. These images can be very dense, requiring high-resolution capabilities, with color as a useful adjunct for labeling, although the original images may be monochromatic.

1.3.2.21   Information Systems

These systems are extensions of the previous two, adding more generalized data to that contained in the others. Those might be considered as subcategories of this application group, and the requirements range from those listed for the others down to the simplest display forms when the information is purely alphanumeric. Therefore, the descriptions of the display requirements and characteristics in the previous sections should suffice for information system applications.

1.3.2.22   Inspection Systems

Inspection systems make up another broad application category, with the requirements covering a wide range, from inspection of simple devices and products to detailed analysis of complex systems. Inspection systems are related to the total measurements procedure, but are unique in the varied types of units that may be used to carry out the operation. Thus, the display application requirements are similar to those for measurements, but may extend beyond that application.

1.3.2.23   Manufacturing

Numerous manufacturing operations still do not use computers for control or to facilitate on-line activity but still require certain types of electronic displays to provide information about the state of the manufacturing operation or what steps are necessary to move to the next stage. This is particularly true for full- or partial-automation operations that are not computer operated or controlled. Much of this information may be presented using A/N readouts, but in some cases larger FPDs or CRTs may be required. Electroluminescent (EL), LCD, and plasma technologies are all in use.

1.3.2.24   Measurements

Measurements have a wide range of visual requirements, from the simple numeric readouts found in portable multimeters to the advanced color displays available in complex analytic systems. Therefore, this application provides a market for all types of display units and technologies, although LCD numerics have captured the portable unit market and CRTs are most prevalent for units requiring advanced display capabilities. Measurement applications are concerned with specific instrumentations, with the display requirements and relevant technologies the same as those for instruments.

1.3.2.25   Message Displays

This is a rather generalized application, in that it may exist with a more specific application. For example, the message display may be used for transportation information such as a large-board display in an airport, or as a scoreboard at a sports stadium. The requirements are generally limited to those associated with A/N modules or panel units, and the relevant technologies may be those associated with flat-panels. However, in some cases, a

CRT unit is preferred, as is the case for the numerous arrival and departure message displays found in airports.

1.3.2.26 Military Applications

Prior to the cessation of the cold war, military applications were the most advanced, and supported the bulk of research and development (R & D) in electronic displays of most types and using most technologies. This application is no longer the cornucopia of funds that it was, but the needs are still there for high-performance displays. Of particular interest are the *helmet-displays* that impose special configuration requirements, and call for numerous developments in projection optics and special formats. These developments may also have some impact on other applications such as multimedia and virtual reality. The main technology has involved the use of small CRTs.

1.3.2.27 Monitoring

This application is very similar to the control and measurement applications in that the same type of information is presented on the display device. It is included here for completeness, and could be considered as part of the applications previously covered. The specification requirements can cover a wide range of performance in that a variety of monitoring functions may be involved, ranging from relatively simple process control to elaborate CAM and CIM installations. Thus, the display portions may be as limited as A/N readouts, or as complex as high-performance CRT systems with full-color and high-resolution capabilities.

1.3.2.28 Multimedia

This is a difficult application to classify as it actually consists of a number of separate applications collected under a single designation. For example, a multimedia installation may be used for animation, art, 3D modeling, presentations, or video editing, to name a few of the possible applications. All of these also exist as separate applications, so the multimedia applications must encompass all of them and the display requirements similarly must cover the entire range of applications. As a result, the multimedia display will be quite complex, and it may be more economical to satisfy only a few of the numerous applications that might be possible. This would lead to a system with less capabilities than might be possible if the best performance were achieved. However, the lower cost should compensate for this reduction in performance if only the most important applications are included. The technologies are both CRT and FPDs, with the former preferred for severe requirements.

1.3.2.29 Navigation

Navigation displays can cover the gamut from simple speedometers to the highly complex head-up and head-down displays found in military aircraft, or the multireadout panels found in commercial aircraft. In addition, the future promises new modes for automotive and aircraft navigation using satellites, with space vehicles promising even more complex systems. This variety of requirements leads to a similar variety of displays with performance requirements specific to each particular application. In the extreme, the requirements include full-color and resolution capabilities, but may be limited to small A/N readouts in the simplest form. The technologies used depend on these requirements and range from those adequate for simple panels to the ones needed for achieving high-resolution color CRTs. This cover essentially all the technologies available for electronic displays, so navigation may be considered as one of the most demanding applications for electronic displays.

1.3.2.30 Presentations
This application generally requires high display performance to achieve the maximum effect, but might also work satisfactorily with only a moderate visual capability. The actual requirements depend on the specific use to which the presentation system is put, and what the source of the presentation information might be. Thus, a full-function video presentation would require high-resolution, full-color CRT or light-valve projection systems, whereas satisfactory slide presentations may be achieved with either LCD projection panels or some of the simpler CRT units that lack the capabilities of the high-resolution systems.

1.3.2.31 Simulation
The visual demands of this application are among the most rigorous of all the applications in use. In the extreme, it requires the capability to generate scenic material of great complexity with the appearance of reality. It is responsible for producing the imagery used in training pilots and others involved in the control of navigational systems. In some cases, the simulations may be quite simple, but for maximum utility, complex systems are needed with all the appurtenances that go with the ability to generate visual material from low to high resolution and with full color.

1.3.2.32 Sports
The display system for sports are those found in the stadiums and racetracks that feature instant playback or multiple off-track presentations. The latter are achieved by supplying CRT monitors in a variety of locations, whereas the former call for large-board displays located centrally on or near the playing field. Scoreboards usually consisting of matrix assemblies of incandescent displays are also found. The instant-playback application can be met by CRT projectors, but light-valve projectors are more satisfactory.

1.3.2.33 Television
Television is the most prevalent application for display systems, with the many millions of TV sets far exceeding the system used for all other individual applications. However, because of their relatively low cost (in

comparison to computers, computer monitors, etc.), TV sets are used for nonentertainment applications as well, and thus quantitatively predominate on the market. Nonentertainment uses of TV systems and/or monitors are covered in later sections on specific applications, this section is limited to entertainment or home TV. Here the CRT-based units are most prevalent by far and therefore CRTs must be considered the leading technology. However, with the advent of high-definition television (HDTV), the demand for large displays has led to an increasing emphasis on improving the performance of FPDs; thus flat-panel technologies should take over, with projection LCD panels in the lead, although plasma panels are still in the running.

1.3.2.34 *Transportation*

This application differs from its associated navigation application in that it is limited to the fixed installations found in flight, bus, and railroad terminals. The emphasis is on large-board displays used for message presentation. In addition, airports use CRT monitors, as noted in the description of the monitors application. However, the use of large-board displays employing electromechanical, magnetic, incandescent, CRT, and similar technologies predominates in transportation applications.

1.3.2.35 *Utilities*

Utilities in the commercial context refer mainly to electricity and gas suppliers, but may also include other sources of necessary products such as water. All of these applications use displays in a similar manner for presenting visual information on the distribution system, and using this information to allow monitoring and control. They might be considered as part of the general control and monitoring application, but this type of application is sufficiently large to stand on its own. A major display format that is used is the large-board displays that show the transmission and distributive systems in a pseudogeographic form. These are supplemented by smaller displays that present portions of the total system to facilitate comprehension.

1.3.2.36 *Virtual Reality*

This is a very new and exciting application. It is difficult to establish exactly what the display requirements might be until more experience is available. However, it does appear that this may be one of the most demanding display applications and will require the highest display performance capabilities to be successful. Small CRTs appear to be most successful in creating convincing displays, but FPDs have the advantage of ease of construction in small sizes.

This completes the brief review and descriptions of applications that require visual presentations to be effective. It is apparent that a wide range of capabilities is required to meet all of them, although any single one may not call for much performance. A number of technologies and product types are used, and any review must cover the totality of visual human factors and display technologies. This is done in Chapters 2 and 3, and a summary of the descriptions and display specifications for those applications covered in this chapter can be found in the tables and text of Chapter 11 for convenience of reference.

1.4   APPLICATION REQUIREMENTS

1.4.1   Introduction

There are a wide range of parameter requirements for the many different applications described in Section 1.3. To achieve the best display performance for each application, one should investigate how the display is used specifically for each application of interest, and then express the requirements for the most appropriate display in terms of its visual parameters as covered in Section 2.2. This is done in detail in the relevant chapters devoted to specific applications. However, it is convenient at this point to briefly discuss a number of applications of particular interest not covered in separate chapters, and present their display specifications to illustrate the process and indicate the importance of establishing properly defined specifications. It is not necessary to deal with those applications that are treated separately in subsequent chapters, but there are still a number of applications that require separate discussion at this point. These are listed in Table 1.2. Several applications that are not listed specifically in this table or elsewhere in the book under separate headings are discussed under a combined or general

TABLE 1.2   Specific Applications

Application	Technology	Display Types
Art	CRT, LCD, EL	CRT, FPD, plotter
Avionics	CRT, LCD	CRT, FPD, projection
Business systems	CRT, EL, plasma, LCD	CRT, FPD, projection
Communication	CRT, LED, LCD	CRT, readout, FPD, large-board
Control/monitoring	CRT, LED, LCD	CRT, readout, FPD, large-board
Desktop publishing	CRT, LCD, EL, plasma	CRT, FPD, plotter
Education	CRT, LCD	CRT, FPD, projection
Geographic systems	CRT, EL, plasma, LCD	CRT, FPD, projection
Imaging systems	CRT, EL, plasma, LCD	CRT, FPD, projection
Inspection	CRT, LED, LCD, VFD	CRT, readout
Manufacturing	CRT, LED, EL, plasma, LCD	CRT, FPD, readout, large-board
Measurements	CRT, LCD, VFD	CRT, readout, FPD
Military	CRT, LCD, EL, plasma	CRT, FPD, projection
Navigation	CRT, LCD, VFD	CRT, FPD
Presentations	CRT, LCD	CRT, FPD, projection
Transportation	CRT, LCD, electromagnetic	CRT, FPD, large-board
Utilities	CRT, LCD, electromagnetic	CRT, FPD, readout, large-board

category as having the same or very similar requirements and therefore not requiring separate treatment. This list covers the most significant applications not included in the subsequent chapters, and should suffice to establish the general approach as well as provide specific parameter requirements for each application listed. First, however, it is necessary to define the parameters that are used to establish the performance capabilities of the hardware so that the numbers will have significance for anyone interested in determining what the expected performance might be as defined by the specifications that are given in terms of these parameters.

1.4.2 Display Parameters

Specifications for displays used in specific applications are best defined in terms of a group of parameters that establish the performance requirements necessary to meet the visual needs of each specific application. A number of these parameters have evolved over the years largely from the lighting, photographic, and cinematic fields, and have been embodied in a group of terms under the general category of photometry, which is defined as "the measurement of quantities associated with light" (*IEEE Standard Dictionary*). The terms are listed in Table 1.3 and are followed by brief definitions and descriptions of each one. These definitions are intended as a partial introduction to the significance of each term (discussed in further detail in Section 2.2). They are detailed as necessary here to explain the significance of the individual application display specifications found in Section 1.4.3 for essentially all the applications listed in Table 1.2.

The parameters and terms listed in Table 1.3 are described in the following list (where the definitions are essentially abbreviated versions of those found in the *IEEE Standard Dictionary*). Luminance (item 4 in the following list) is probably the most commonly used parameter to define the performance of a visual display. The definition given in the following list is too complicated to be useful, and for the purposes of application requirements it is sufficient to consider luminance as the parameter measured by a luminance meter. It is essentially a measure of the light output of a luminous source, and is the instrumental equivalent of the psychological term, brightness, which is frequently used to mean luminence. This usage is to be frowned on [2]. There is also some further confusion in the various terms that are used to define the luminance values. The international term is cd/m^2

TABLE 1.3   Photometric Parameters and Terms

Lumen	Candela (cd)	Steradian
Luminance (L)	Photometric brightness	Luminous flus (F)
Luminous intensity (I)	Illuminance (E)	Illumination (E)
Color (C_n)	Contrast (C)	Contrast ratio (C_R)

(candela per square meter), whereas foot-lamberts (fL) is still used in the United States. The conversions from foot-lamberts to lux and the reverse are given in Table 2.3.

1. *Lumen.* The lumen is the basic unit luminous flux, and is defined as the flux through a unit solid angle (steradian) from a point source of 1 cd emitting equally in all directions.

2. *Candela.* The candela is the basic unit of light intensity, defined more fully in Section 2.2.2.

3. *Steradian.* This is a standard geometric term, defined as a unit solid angle.

4. *Luminance.* Luminance is officially defined as the quotient at a point of a surface in a given direction of an infinitesimal element of the surface containing the point under consideration, by the orthogonally projected area of the element on a plane perpendicular to the given direction.

5. *Photometric Brightness.* This term is identical to *luminance,* and is infrequently used.

6. *Luminous Flux.* The luminous flux emanating from 1-cd point source is 1 lm within the unit solid angle, and 1 lm is defined as the flux on a unit surface with all points at a distance of one unit from the source. This is illustrated in Figure 1.1.

7. *Luminous Intensity.* Luminous intensity is the solid-angle flux density in any given direction, coming from a point source with the value of one candela (cd).

8. *Illuminance.* This term is defined as the density of luminous flux incident on a surface. The international term is lux, and the U.S. equivalent term is footcandle (ft-C) requiring conversion factors as shown in Table 2.3.

9. *Illumination.* This term is the exact equivalent of illuminance and is included for information.

10. *Color.* Color refers only to the number of different colors that can be generated.

11. *Contrast.* Contrast is an important measure of the legibility of a display, and is defined as the ratio of (a) the total luminance at a point minus the surround luminance to (b) the surround luminance. The ratio of this difference to the higher luminance may also be used, leading to a fraction instead of a positive number. This was a common formulation in the past, but it has been superseded by contrast ratio, defined next.

12. *Contrast Ratio.* As noted above, this phrase is preferred to *contrast,* and is defined as the ratio of the total luminance to the surround luminance.

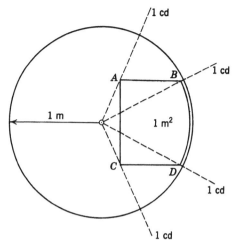

Photometric Terms

Term	MKS Unit[a]	Definition
Luminous intensity (I)	Lumen / W (cd)	Solid-angle flux density
Illuminance (E)	Lumen / m² (lux)	Flux density on surface
Luminance (L)	Lumen / (W · m²) (nits)	Flux density per unit area
Luminous flux (F)	Lumen	Time Rate of flow of light
Candela (cd)	Candela	Unit of luminous Intensity

Figure 1.1 Relation among candelas, lumens, and lux. After Kaufman [1]. Reproduced from the *Lighting Handbook*, 4th ed. Published by the Illuminating Engineering Society of North America, 120 Wall St., 17th Floor, New York, NY 10005—212-248-5000.

Further information on these parameters may be found in Section 2.2. Additional nonphotmetric parameters are resolution, speed, viewing area, triads, and matrix. These terms are largely self-explanatory except for resolution triads and matrix. *Resolution* refers to the number of discernible visual elements, *triads* are concerned with the three primary colors used in a color display; and *matrix* with an assemblage of individual elements in *X–Y* format.

1.4.3   Application Display Specification

1.4.3.1   Art

This is a relatively complex type of application, with the visual results ranging from the simplest cartoon to the maximum complexity of fine art. The user is an artist more accustomed to materials such as brushes, palette knives, and a wide range of oil, pastel, and watercolor media. The output when using these media can cover the gamut in colors and line thickness. Thus, if the

TABLE 1.4 Display Specification for Art Applications

Parameter	Value	Technology	
		Input	Output
Luminance (cd/m^2)	3.0–30	NA	CRT, FPD, plotter, slide
Contrast ratio (number)	10–50	NA	NA
Resolution (TV lines)	1000–3000	Light pen, mouse, trackball	CRT, FPD, plotter, slide
Speed (in./s)	1–5	Light pen, mouse, trackball	CRT, FPD, plotter, slide
Colors (number)	256–16.7 million	NA	CRT, FPD, plotter, slide
Viewing area ($H \times W$, in.)	5 × 5–30 × 20	12 × 12–30 × 20	CRT, FPD, plotter, slide
Triads (number/line)	3000–9000	NA	CRT, FPD
Triads pitch (in.)	0.1–0.3	NA	CRT, FPD
Matrix (columns × rows)	512–2000	NA	FPD
Matrix element (in.)	0.1–0.3	NA	FPD

electronically produced image is to match that possible with manual methods, it must be capable of at least approximating that produced by more conventional means. It is difficult to quantify what the actual parameter values should be because the manual results are so varied. To ensure that electronic "art" is capable of approximating the best results from manually produced fine art, it is desirable to achieve the maximum in performance, particularly if hard copy is to be generated by a color plotter. A specification is shown in Table 1.4, with the parameter values chosen to achieve the best possible performance. The form of the system might be as shown in Figure 1.2, which is a generalized block diagram of the system; a simpler one is shown in

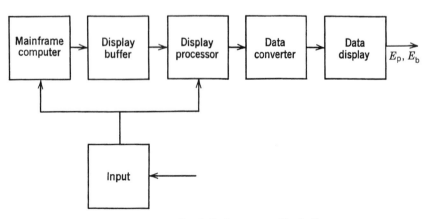

Figure 1.2 Generalized display system block diagram.

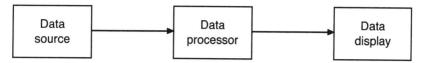

Figure 1.3 Simplified display system block diagram.

Figure 1.3. In addition, of these, only the input and output blocks are of immediate concern to the user, as—with the proliferation and variety of computers—one can surely be found to fulfil all the data processing functions. In that case, at least some of its performance parameters are of interest, but this aspect of total display system performance is covered in Chapter 2 on greater detail. Therefore, Table 1.4 lists only those parameters that define primarily the performance capabilities of the input and output devices, and the technologies available for implementing these devices. This display specification is presented here as an example, but it does not apply accurately to the art application. It is repeated in the Chapter 11 tables under the fine art label for information and reference, along with similar application descriptions and display specifications for the rest of the applications listed in Tables 1.1 and 1.2. These descriptions and specifications may be referred to for information on application display requirements, along with other descriptions and applications found in Chapters 4–10, which deal in more detail and from several points of view with several of the most important specific application groups. This approach leads to some apparent redundancies, but is necessary to fully cover these important application groups without requiring the reader to refer to other chapters for complete coverage. An example of an important group is graphics arts, discussed in Chapter 5, which includes a number of the applications listed in Tables 1.1 and 1.2 focusing on the graphics art aspects of their display requirements. This approach is also used in the other chapters dealing with specific generalized applications, and accounts for some unavoidable redundancy albeit more apparent than real, if full detail is desired. With this short preliminary explanation, we may proceed to the following chapters with a better understanding of the structure and format used in this volume.

REFERENCES

1. Kaufman, J. E., Ed., *IES Lighting Handbook*. 4th ed. Illuminating Engineering Society of North America, New York, 1966.
2. Sherr, Sol, *Electronic Displays*, 2nd ed., Wiley, New York, 1993.

2

HUMAN FACTORS CONSIDERATIONS AND TECHNOLOGY REVIEW

2.1 INTRODUCTION

Although the cathode ray tube (CRT) remains the most common type of display device, and in addition to the proliferation of laptop, notebook, subnotebook, and various other portable computers such as the Newton, as well as a host of other portable, battery-operated equipment, there has been an exponential growth in the sale and use of other display technologies more compatible with the requirements of handheld units. This has not led to the demise of equipment using CRTs as the major display device, in particular for television applications, for reasons covered in some detail in the CRT section of this chapter and elsewhere in this volume, as appropriate. However, the advent of high-definition television (HDTV) has led to increased demand for FPDs to meet the display requirements of this technique, and this application makes up a major portion of the market. Furthermore, in spite of attempts, with some success, to produce flat-panel versions of CRT devices that have been incorporated into portable units, in general the other flat-panel technologies have essentially superseded all of the CRT-based approaches to flat-panel displays. This is not to state unequivocally that some new approaches to vacuum-tube displays, such as the use of multiple emitters described in the CRT section, may not succeed in competing with the various other technologies that have been most successful in providing displays for equipment requiring flat-panel displays. Therefore, new CRT-based technologies should not be ignored.

Foremost among the many technologies that have been investigated as a means of developing flat-panel displays (FPDs) that meet the requirements of the various types of portable equipment are those listed in Table 2.1*a*. Although a number of other technologies have been investigated and devel-

TABLE 2.1   Flat-Panel Display Technologies

Technology	Type	Drive
a. Major FPD Technologies[a]		
EL	Light-emitting	Ac, dc
LED	Light-emitting	Dc pulse
LC	Passive	Ac pulse
Plasma	Light-emitting	Ac, dc
Vacuum fluorescent	Light-emitting	Ac pulse
b. Other FPD Technologies		
CRT	Light-emitting	Ac pulse
Electrochemical	Reflective	Dc pulse
Electrochromic	Passive	Dc
Electrophoretic	Passive	Dc
Electromagnetic	Reflective	Dc pulse
Field emission	Light-emitting	Ac pulse
Incandescent	Light-emitting	Ac
Laser	Light-emitting	Ac pulse
Particle	Reflective	Dc

[a]Other technologies that have been tried with only limited success or are too new for a final evaluation to be made are shown in part *b* of this table.

oped to some extent, and others show varying degrees of promise, those shown in Table 2.1*a* are the most successful.

A number of other technologies in addition to those listed in Table 2.1 have been tried and at least for the moment forgotten, such as electrostatically operated reflective elements and flipping spheres, but these and other similar approaches are not covered here.

2.2   HUMAN FACTORS CONSIDERATIONS

2.2.1   Introduction

Human factors is the generic term for all psychological and physiological characteristics of the humans involved in the use of the systems and the parameters that define the system performance in terms that relate it to these human characteristics. The human visual system is of primary importance in establishing the performance parameters and capabilities and limitations. The actual system performance requirements can be established in terms of these parameters, and the expected limitations described, for the specific applications asdetermined within these limits of known human visual and recognition capabilities. It is very important to understand what can and cannot be achieved within the limits of human capabilities to ensure that excessive requirements are not imposed on the display system.

Before examining the visual parameters it is desirable to describe the human visual system to some extent to establish a basis for the choice and range of values of these parameters. The simplified diagram of the eye shown in Figure 2.1 represents this organ from its external beginning to the point at which the visual inputs are converted into electrical signals, which occurs at the retina shown at the lower end of the diagram. As the light quanta pass through the cornea, which is a tough, transparent tissue without blood vessels, they are refracted by the corneal surface and pass on to the lens surface, where they are further refracted. The iris acts as a shutter (similar to the F stop or aperture in a camera lens) to control the amount of light that

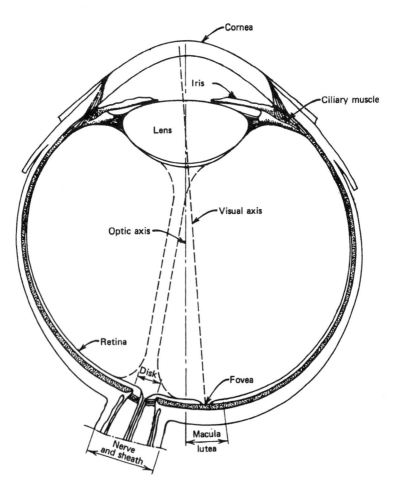

Figure 2.1 Diagram of cross section of the human eye. After Cornsweet [1, p. 40], by permission of Academic Press.

reaches the lens. Up to this point the eye is similar to a camera, except that the ocular lens is flexible and its curvature can be changed by means of muscles attached to the eyeball. This allows the focal length of the lens to be changed and accounts for our ability to retain sharpness of vision over a range of distances, at least while we are young. Detailed discussion of the various visual problems encountered by humans are beyond the scope of this volume, and the interested reader is referred to other sources [1] for such details. Suffice it to note that near- and farsightedness are common difficulties, usually corrected by spectacles, that have some impact on the ability of human observers to extract maximum information from a display and play a role in determining usable parameter values.

Continuing with the discussion of the various elements of the eye, the next one of major importance is the retina, which consists of a very large number of receptors that are light-sensitive and convert the light that impinges on the retina into electrical signals. There are two types of light-sensitive elements in the human eye, which are termed *rods* and *cones*; there are about 100 million rods and 7 million cones. They differ in their response characteristics; the rods respond primarily to low-level light without color differentiation, whereas the cones (of which there are three types) respond primarily to color, with each type sensitive to one of the three primary colors. The rods are the most sensitive, capable of being activated by as little as 100 quanta, but the cones provide the best resolution, being tightly packed at the center of the eye, in the region termed the *fovea*. The rods are more diffusely located, spread over the entire retinal area, which explains why visual resolution is always worse at low light levels, as the rods are not activated at these levels.

The outputs from the rods and cones in the retina are series of pulses that are sent through the optic nerve bundle into the brain, where they are interpreted into the visual image. However, this part of the visual process is the subject of physiology and psychology, and this volume is limited to those aspects of the latter that can be determined by means of experiments and measurements that can be made without penetrating into the brain. The next step in our examination is to discuss the various parameters used to define and describe visual performance, beginning with *photometry*, or the measurement of photometric events, and proceeding to the visual parameters that establish the aspects of visual performance of interest in the application to be covered.

2.2.2 Photometric and Visual Parameters

2.2.2.1 *Photometric*
Photometry is defined in Section 1.4.2 as "the measurement of quantities associated with light" or more specifically as "the techniques for the measurement of luminous flux and related quantities." These quantities are listed in Table 1.3, and are repeated here in Table 2.2. All of these terms are defined

TABLE 2.2   Photometric Parameters and Terms

Lumen	Candela (cd)	Steradian
Luminance (L)	Photometric brightness	Luminous flux (F)
Luminous intensity (I)	Illuminance (E)	Illumination (E)
Color (C_n)	Contrast (C)	Contrast ratio C_R

in Section 1.4.2. However, more extended definitions of some of these terms and mathematical statements are given next, as well as some conversion factors for changing from meter-based to foot-based parameters.

1. *Lumen.* The meaning of the lumen is illustrated in Figure 2.2, which shows the flux through a unit solid angle (steradian) from a point source of 1 cd emitting equally in all directions.
2. *Candela.* The basic unit of light intensity, is the luminous intensity of a blackbody radiator at the temperature of solidification of platinum with a projected area of $\frac{1}{60}$ cm.
3. *Luminance.* The various terms used to define luminance values are confusing because the international (metric) term *lux* (lumens/m^2) as well as the customary U.S. term, *foot-lambert* (ft-L), are still used in the United States. Therefore, conversions from one term to the other must be considered, and the conversions from ft-L to lux and the reverse are given in Table 2.3 with other conversion factors of interest.

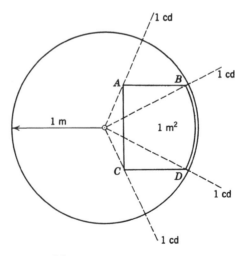

Figure 2.2 After Kaufman [2]. Reproduced from the *Lighting Handbook*, 8th ed. Published by the Illuminating Engineering Society of North America, 120 Wall St., 17th Floor, New York, NY 10005—212-248-5000.

TABLE 2.3   Photometric Conversion Factors

| | Multiply Number of | | | | |
To Obtain	cd/ft^2	cd/m^2	ft-L	lux	ft-C
			by		
Foot-lambert (ft-L) or Fl	3.142	0.2919	1.0	—	—
Candela/meter^2 (cd/m^2)	10.764	1.0	3.426	—	—
Lumen/meter^2 (lux)	—	—	—	1	10.76
Lumen/ft^2 (ft-C)	—	—	—	0.0929	1.0

4. *Luminous Flux.* Referring again to Figure 2.2, 1 lm is defined as the flux on a unit surface with all points at a distance of one unit from the source.

5. *Luminous Intensity.* Luminous intensity is illustrated in Figure 2.2 as coming from a point source with the value of one candela (cd).

6. *Illuminance.* As shown in Figure 2.2, for the one meter (1 m) distance, the flux is given by

$$E = 1 \, \text{lm/m}^2 \, (\text{lux}) \tag{2.1}$$

and if 1 ft is the distance, then

$$E = 1 \, \text{lm/ft}^2 \, (\text{ft-C}) \tag{2.2}$$

Therefore, as is shown in Table 2.3, to convert from lux to ft-C, multiply lux by the ratio between m^2 and ft^2 or 0.0929, and multiply ft-C by 10.76 for the opposite.

7. *Contrast.* Contrast is defined mathematically by

$$C = \frac{L_T - L_S}{L_S} \tag{2.3}$$

The ratio of this difference to the higher luminance may also be used, leading to a fraction instead of a positive number. This was a common formulation in the past, but both this ratio and Equation 2.3 have been superseded by contrast ratio, defined next.

8. *Contrast Ratio.* Contrast ratio is defined mathematically by

$$C_R = \frac{L_T}{L_L} \tag{2.4}$$

The conversion factors for the meter-based and foot-based versions of these terms are given in Table 2.3, and continue to be used although to a lesser extent than previously.

2.2.2.2   *Visual*

In addition to the photometric parameters, there are several visual parameters of interest: resolution, visual acuity, legibility and flicker. These parameters are described next, with legibility as potentially the most meaningful as an evaluation technique for determining the performance capabilities of electronic displays.

Resolution   Resolution is a difficult parameter to specify because of the many ways in which it is defined. The most common types are shrinking-raster and television-limiting, resolution and these are the only types discussed in the following sections.

SHRINKING RASTER RESOLUTION   This type of resolution measurement may best be understood by reference to Figure 2.3, which shows a raster of lines that corresponds to the movement of the electron beam when it is generated by a succession of sweeps across the face of a CRT. These are similar to those generated by a TV sweep generator, but differ in that the number of lines can be controlled and the height of the selected number on the CRT screen can be altered. This procedure is used to generate a specific number of lines and then adjust the height of the image to some selected value at which the raster lines begin to merge. The resolution is then defined by the ratio of the number of lines to the height of the raster, or

$$R = X \text{ lines}/Y \text{ cm} \tag{2.5}$$

This is a convenient way to measure and specify resolution, but it should be noted that it differs by a factor of almost 2 in terms of the number specified from that found by using the TV-limiting technique, discussed next. In addition, it is not the same as that obtained from a direct measurement of

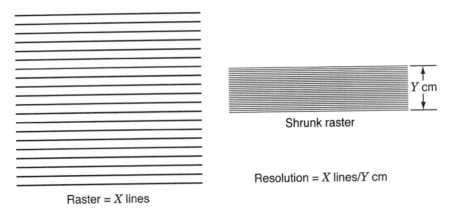

Raster = X lines

Figure 2.3   Shrinking-raster measurement.

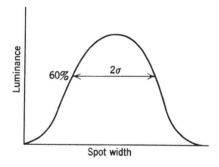

Figure 2.4 Significance of 2σ.

the width of a single line or CRT spot, because the CRT spot profile is Gaussian, as shown in Figure 2.4, and the spot width depends on where the measurement is made in this profile. Conversion factors for these and other measurement techniques are shown in Table 2.4.

TELEVISION-LIMITING RESOLUTION   This technique for making resolution measurements is frequently used because of the wide prevalence of TV systems as the electronic display means. It is based on the use of a test chart such as that shown in Figure 2.5, which is the one issued by the Electronics Industries Association (EIA) and is standard for this measurement. A full television-type display system must be used, with a TV camera to pick up the chart image and convert it into the equivalent image on the face of the CRT being measured. The resolution is then the number shown on the chart at the point where the lines merge, and in that respect the technique is similar to that used for the shrinking-raster method. It should be noted that both of these measurement techniques are subject to human error, although an individual can usually repeat the readings with reasonable consistency. However, both are simple to implement and are generally used to specify this parameter.

MODULATION TRANSFER FUNCTION   A more reliable technique is the modulation transfer function (MTF), which is the most accurate and reliable of all

TABLE 2.4   Conversion Factors for Resolution

To Convert from	TV-Limiting	10% MTF	Shrinking-Raster	50% Amplitude	50% MTF
TV-limiting	1.0	0.8	0.59	0.5	0.44
10% MTF	1.25	1.0	0.74	0.62	0.55
Shrinking-raster	1.7	1.36	1.0	0.85	0.75
50% amplitude	1.0	1.6	1.17	1.0	0.88
50% MTF	2.26	1.82	1.33	1.14	1.0

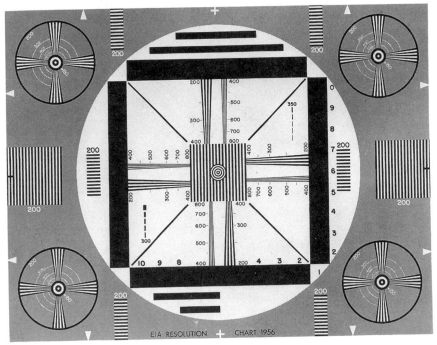

Figure 2.5   EIA resolution chart.

the possible techniques, but requires rather expensive equipment and may not be used by the manufacturer of the display being measured. However, the conversion factors among all four of the techniques mentioned are listed in Table 2.4 to give the reader some idea of the relationship. Of course, if it is possible to obtain the MTF for the device or system, it should be considered the most accurate and desirable of the possible measures of display resolution.

SPOT SIZE   This has been a relatively unusual way to specify resolution other than for a CRT alone, but is becoming more common for display systems with the advent of HDTV where spot size in relation to display size is a significant factor in setting limits on the maximum number of scanning lines that can be viewed on a CRT display.

The 50% amplitude signifies the place where the spot width is measured. The other designations correspond to the descriptions given previously. The resolution chart shown in Figure 2.5 is used for the TV-limiting measurement, and the shrinking-raster measurement uses the set of scanning lines as shown in Figure 2.3. The MTF measurements require the proper equipment, such as a slit analyzer system.

Visual Acuity This is next in the list of nonphotometric visual parameters, as it is closely related to resolution. Specifically, it is defined as the ability to discriminate fine detail in the field of view, in relation to normal acuity at a standard distance D', and by the equation

$$\text{Visual acuity} = \frac{D'}{D} \qquad (2.6)$$

where D is the distance at which the minimum discernible test object subtends 1 min of arc. One minute of arc at the eye is the accepted limit of its resolving power, and this leads to another definition for resolution as the reciprocal of the angular separation between two elements of a test pattern when they are just detectable as separate elements. This may be considered the best resolution of which the human visual system is capable, and anything more than this is unnecessary. It is important to bear this limitation in mind as higher resolution may be expensive to achieve and add nothing to the general quality or legibility of the display.

Legibility Legibility is probably the most readily convertible parameter into a direct and fully meaningful determination of how well the display device or system will function in the actual application environment. Briefly, legibility is determined by measuring how well the information presented on the display surface can be communicated to the user under the actual conditions imposed by the application. Tests are performed, such as how many characters can be accurately recognized under these conditions, or by variants of this procedure. It is important to correctly choose the types of identifications to be made if the tests are to be fully meaningful, and this is a task for a human factors expert who is also familiar with the application requirements.

Flicker Finally in this survey of nonphotometric visual parameters there is flicker. This is an extremely subjective parameter as one person's flicker may be another's acceptable display. Flicker results from the fact that CRT-based and other nonretentive displays must be refreshed at some repetition rate for the display information to be constantly visible. This is because the electron-beam energy of CRT phosphors and the electrical signals of other light-emitting elements such as thin-film electroluminescence (TFEL) and plasma are converted into visible images. The phosphors used for CRTs are subject to decay times ranging from very short to very long, but must in general be short enough so that the visual data are not smeared. Unfortunately, the human visual system is capable of detecting the on–off characteristics of such a presentation if the repetition rate is below some rate, as noted previously, and defined as the critical flicker frequency (CFF). The mathematical expression for the CFF is known as the Ferry–Porter law, expressed since 1987 by

$$CFF = m + n\{[\ln E(f)_0]\} \tag{2.7}$$

where m and n are constants dependent on viewing angle, and

$$E(f)_0 = ae^{bf} = \text{amplitude of fundamental refresh frequency}$$

and a and b are constants dependent on display size. Curves of CFF versus luminance as predicted by this equation are given in Figure 2.6, and indicate that this parameter is subject to individual characteristics and the luminance level, or—to be more specifically physical—the brightness of the display. A usual value for an acceptable refresh frequency at normal luminance levels is in the vicinity of 50 to 60 Hz, so that the normal rate of 60 Hz, used in the United States and other countries with a 60-Hz power-line frequency, is usually adequate to avoid flicker. However, it has been found that refresh rates as high as 70 Hz may be required in certain situations, and the 50-Hz rate used in Europe and other countries with a 50-Hz power-line frequency is wholly inadequate, especially when an interlaced frame rate is used.

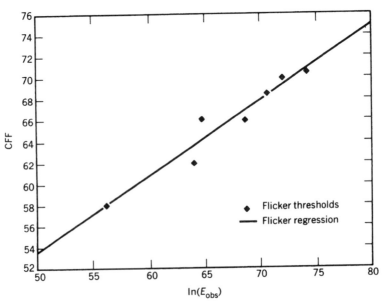

Figure 2.6 Critical flicker frequency (CFF) versus natural log of absolute amplitude. After Farrell et al. [3a], by permission of Society for Information Display (SID).

2.3 MAJOR TECHNOLOGIES

2.3.1 Introduction

As noted in Table 2.1*a*, the most important technologies for FPDs are LCDs, VFDs, dc plasma, EL, LEDs, and ac plasma, more or less in that order, with CRTs relegated to secondary status along with the other technologies shown in Table 2.1*b*. However, CRTs remain probably the most important technologies for many applications, especially those that require the highest visual quality and need not be portable. In addition, to a considerable extent, the majority of FPDs use the LC technology with the next two mainly used for numeric and A/N displays containing a limited number of characters. EL and ac plasma technologies have found some application to larger FPDs with both A/N and graphics displays. As noted previously, LCDs have tended to take over the portable applications for instruments and computers, but vacuum-fluorescent displays (VFDs) and dc plasma are managing to hold their own for readouts in applications where ac power is available. LEDs are used primarily as readouts when large character sizes are desired. Therefore, although active- and passive-matrix LCDs are making rapid progress in providing excellent quality at fairly reasonable prices, CRT-based displays remain in the lead where low power requirements are not an important factor.

EL and ac plasma can compete in many respects with LCD and CRT displays where good legibility and image quality are important, but the difficulty in achieving acceptable color at reasonable cost and complexity remains a deterrent. However, development work continues, especially in color EL, so that the future may bring a change. In any event, all the technologies mentioned above are given detailed treatment in the following sections, beginning with the nonFPD CRT types, and followed by the major FPD technologies in alphabetical order.

2.3.2 CRT

2.3.2.1 General Description
CRTs warrant primary consideration because the devices using this technology remain in the forefront of the displays used for a wide variety of applications described in Chapter 1 and later chapters. The discussion here concentrates on the technological factors and the performance capabilities and limitations that result from these factors. The analyses are fairly complete given the limitations in space. First and perhaps foremost, it should be noted that this technology is probably the oldest among the display technologies, dating back to the original invention by Braun at the end of the nineteenth century, and remarkably similar in basic concept to the original invention. Briefly, a CRT is defined as "an electron tube, the performance of which depends on the formation and control of one or more electron beams"

(*IEEE Dictionary*). Somewhat more specifically, it consists of a thermally heated cathode that emits a beam of electrons, which in turn is controlled in intensity by a grid, and focused into a desired cross section by a set of focusing electrodes. This portion of the tube is known as the electron gun, and is shown schematically in Figure 2.7. It should be noted that this assembly is contained within an evacuated envelope and the focused electron beam is the cathode ray part of the CRT.

Following and surrounding the beam in some fashion are means for deflecting the beam orthogonally to cause the electrons to strike a specific range of locations on a surface. This surface is usually coated with a luminescent material that can be made to emit light when energy-carrying electrons strike it. The energy is imparted to the electrons by means of an accelerating voltage that is placed on the sealed and evacuated bottle that contains the electron gun. This entire assembly is shown in Figure 2.7, where only one means for deflecting the beam is included, although two means—electrostatic plates or magnetic coils—are available. The first consists of orthogonal plates used to cause electrostatic deflection, and the magnetic coil results in magnetic deflection when the proper signals are applied. Similarly, there are two means for focusing the beam, again electrostatic or magnetic, where the structure for electrostatic focus is shown schematically and is surrounded by the accelerating electrode. The location for the magnetic focus coil is shown later in the section on magnetic focus (see also Fig. 2.8).

CRTs appear in numerous forms where the output may be in one or many colors, and the number of beams may range from one to thousands. For example, the monochromatic versions use a single phosphor, whereas color tubes use up to three different phosphors, and some flat-panel units may have hundreds or more different beams. Finally, it is appropriate to examine each technique used in CRTs and briefly referred to above. These are beam production and emission, magnitude control, focusing, deflection, acceleration, and beam-to-light conversion.

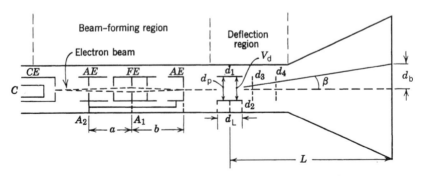

Figure 2.7 Electrostatic deflection and focus CRT. From Sherr [3b] by permission of John Wiley & Sons, Inc.

2.3.2.2   CRT Techniques

Beam-Forming Region   This is the region responsible for producing the electron beam and then accelerating it and controlling its density. It is the CRT gun, also referred to as the *triode* because it consists of three elements, that is the electron-producing element or cathode, the beam control element or grid, and the beam accelerating element or first anode. The electron beam may be produced in a variety of ways, but the most common one is to have an electrode made of a material such as a mixture of strontium and barium carbonate on a nickel base, that is capable of producing a cloud of electrons when it is heated, wound around an element that can be heated to an appropriate temperature when an electric current is passed through it. This latter element is known as the *filament* or *heater*, whereas the electron source is called a *thermionic cathode*, and the combination is termed an *indirectly heated cathode*. The cathode may also be heated directly, but this is not the rule for CRTs and need not concern us further. When the cathode is heated to the proper temperature for electron emission, a cloud of electrons appears in the vicinity of the cathode and produces a space charge in the absence of a potential that can accelerate the electrons away from the cathode vicinity. The cathode is followed by the control grid and accelerating anode, and the function of the control grid and accelerating voltage is to remove this cloud from the vicinity of the cathode so that it can be focused and deflected. The beam current can be expressed in terms of the voltages on the control and accelerating grids as

$$i_k = 3E_d^{3.5}/E_{cc}^2 \tag{2.8}$$

where i_k is given in microamperes and the other elements are in volts.

Beam-Shaping Region   The beam next passes through the focusing region, which may consist of an electrostatic lens made up of a focus grid as shown in Figure 2.7, or a magnetic lens made up of a focusing coil as shown in Figure 2.8. Dealing first with the electrostatic focus technique, the focusing elements consist of the plates shown in Figure 2.7, which act as an electrostatic lens, and cause the electrons that have been accelerated by the voltage applied to the accelerating anode to focus into a narrow beam. The effect is analogous to what happens in visual optics with an optical lens, and similar equations apply as is demonstrated by the equations listed in Table 2.5. The parallels are quite close, which should help photography enthusiasts better understand electron optics. In any event, a background in optics is quite useful for establishing the equivalence, and optics texts may be consulted to further characterize this relationship. However, the information contained in Table 2.5 should be adequate for achieving an understanding of the role of electron optics in CRTs.

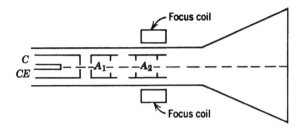

Figure 2.8   Magnetic focus CRT. From Sherr [3b] by permission of John Wiley & Sons, Inc.

In actual systems, both optical and electrostatic, most lenses are more complex than those covered by the equations given in Table 2.5. However, these equations are sufficient to illustrate the parallelism between the two types. In addition, both the optical and electrostatic or electron optical focusing lens lead to a number of aberrations as listed in Table 2.6.

The equivalence of the two types is obvious, as is the fact that the electron beam may take on a variety of shapes and has a finite dimension. The distribution is Gaussian, and the width is measured at the 60% point as shown in Figure 2.4. Thus, when the spot size is given in the specification, it may be assumed to mean the width at this point of the spot, and there will be some overlap. In addition, the resolution in TV lines corresponds approximately to spot size measurement at the 10% point so that the overlap is even larger. Therefore, it is wise to take care when converting these numbers into the actual legibility of the total display.

The focusing technique described above is exclusively electrostatic, which is also termed *electrooptic* to highlight these equivalences. However, there is

TABLE 2.5   Properties of Thin Lenses

Optical Lenses	Electron Optical Lenses
	Snell's Law
$n_1 \sin \theta_1 = n_2 \sin \theta_2$	$\sqrt{V_1} \sin \theta_1 = \sqrt{V_2} \sin \theta_2$
(where n = index of refraction)	V = potential
	Geometry
$(1/p) + (1/q) = 1/f$	$(1/P) + (1/Q) = 1/F$
(where f = focal length	[where f = focal length = $4V/(E_1 - E_2)$]
= $\{(n - 1)[(1/R_1) + (1/R_2)]\}$;	$m = (q/p)\sqrt{F_1/F_2}$
$m = h_2/h_1 = q/p$	

Source: Keller [8].

TABLE 2.6 Aberrations

Type	Optical	Electron Optical
Chromatic	Rays of different wavelengths have different focal lengths	Electrons of different energies have different focal lengths
Spherical	Rays parallel to the axis have a focal length that depends on the distance from the axis	Electron rays parallel to the axis have a focal length that depends on the distance from the axis
Astigmatism	Off-axis rays have a focal length dependent on their angle	Off-axis electrons have a focal length dependent on their angle
Coma	A round object is focused into a tear-shaped image	A round crossover is focused into a tear-shaped spot
Curvature	Rectangular grids become curved, pincushioned, or barreled	Rectangular grids become curved, pincushioned, or barreled

another means for focusing the beam referred to previously, namely the electromagnetic, usually termed *magnetic*. This technique employs a coil placed around the neck of the CRT, as shown in Figure 2.8, and the beam focusing is accomplished by varying the current through the coil. This is because when an electron enters a uniform magnetic field, it follows a helical path and the pitch of the helix is controlled by the magnetic field, as is the radius of the electron path. The applicable equations are somewhat too complex to be of interest to users, but it is important to note that somewhat better focusing can be achieved with magnetic as opposed to electrostatic focusing. Therefore, although electrostatically focused CRTs are most commonly used for the majority of the applications, magnetic focusing may be found when the smallest spot size is desired. It should also be noted that higher luminances may be achieved with magnetically focused CRTs.

Beam-Deflection Region As with focusing, beam deflection may be achieved either by electrostatic or magnetic means. However, whereas electrostatic focusing is most common, the reverse is true for deflection, where the majority of CRT systems use magnetic deflection. Beginning with electrostatic deflection to maintain uniformity of treatment, the deflection means are the orthogonal plates shown in Figure 2.7, and more schematically in Figure 2.9, where only one pair of plates is shown for clarity. In this case the deflection is given by

$$y_d = \frac{lb}{2n} \frac{V_d}{V_a} \qquad (2.9)$$

The other pair of plates control the motion in the orthogonal direction, and follow the same equation. It can be seen from Equation 2.9 that the amount of deflection is dependent on the level of the deflection voltage, where V_a is

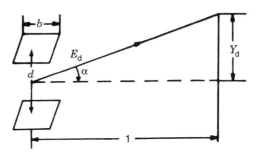

Figure 2.9 Electrostatic deflection. After Sherr [4], by permission of Marcel Dekker.

the initial accelerating voltage and is fixed for any monitor design. The two deflection voltages are each the difference between the pair of plates, and the combination can move the spot over the surface of the CRT faceplate. This technique is quite effective in achieving rapid deflection, because there is only a time constant consisting of the plate capacitance and wiring resistance, so in this respect it is superior to the magnetic deflection technique discussed next. However, electrostatic deflection may require rather high deflection voltages for wide angles of deflection, and the luminance is less than for a magnetic deflection CRT, so the latter is preferred for most applications.

Magnetic deflection is illustrated in Figure 2.10, where again a single axis is shown, but the actual magnetic deflection coil or yoke consists of two orthogonal coils that control the two axes as in the case of electrostatic

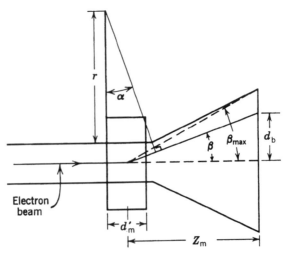

Figure 2.10 Magnetic deflection CRT. After Spangenberg [5], by permission of McGraw-Hill.

deflection plates. The relevant equation is

$$d_b = \frac{(k_m L_y)^{1/2} I_y}{(2V_r)^{1/2}} \tag{2.10}$$

where L is the yoke inductance and I is the current through the yoke, and it is assumed that the deflection angle is small enough so that the sine of the angle is equal to the deflection angle, which in turn is equal to the deflection. Thus, the deflection is controlled by the acceleration voltage, the yoke inductance, and the yoke current, and since the first two are fixed in any monitor design, the deflection is basically proportional to the current. The x and y deflections are similarly controlled by the current through each yoke element, and the resultant beam deflection moves the spot over the total screen surface. It should be noted that the length of the tube is controlled by the deflection angle, with the smaller deflection angle resulting in the smaller CRT length. This is shown in Table 2.7 for angles up to 110° and illustrates why wide-angle CRTs are preferred and also illustrates the advantage of magnetic over electrostatic because shorter lengths are possible with the former. There are also other reasons why electrostatic CRTs are so rarely used that they are found only in some oscilloscopes.

As indicated in Figure 2.10, the yoke is placed around the tube neck and is in the shape of a coil with a gap through which the electron beam passes. The deflection center is set by the yoke location, and the deflection amount is controlled by yoke sensitivity and maximum deflection angle. The advantage of magnetic over electrostatic deflection is that the first is proportional to $V^{1/2}$ whereas the second is proportional to V. Therefore, smaller currents may be used, and in addition, current is easier to generate than voltage. As a result, electrostatically deflected CRTs are used only for special-purpose applications, such as oscilloscopes, or where speed of deflection is paramount.

Light-Production Region   This region consists primarily of the phosphor on the faceplate and the ultor or high-voltage acceleration electrode. The

TABLE 2.7   **CRT Bottle: Deflection Angle versus Length**

Deflection Angle (°)	CRT Length (cm)
35	70.6
45	56.1
60	43.2
70	37.3
90	29.2
110	23.6

Source: Herold [9].

voltage applied to this electrode is quite high, as much as 20 kV or more for color CRTs, and in the thousands for monochromatic units. Its function is to accelerate the beam electrons to ensure that they achieve enough energy to excite the phosphor into emitting light. The color of the emitted light is determined by the phosphor or phosphors used, and three separate phosphors are necessary to achieve a full color gamut. The intensity of the light is controlled by both the magnitude of the acceleration voltage and the beam density. The former is set by the monitor design and the latter is controlled by what is usually termed the *brightness control*, which is something of a misnomer because of the confusion between luminance and brightness. The luminance is given in terms of accelerating voltage and current by

$$L = k_b I_b \frac{V_b^n}{A} \tag{2.11}$$

where k_b = proportionality factor shown in Table 2.8 in lumens per beam
watt = $\eta\gamma$
γ = lumens per radiated watt
η = phosphor efficiency
A = area of phosphor surface

These terms are listed in Table 2.8 for a number of popular phosphors, and full details on these and other registered phosphors may be found in reports provided by CRT manufacturers as well as other related information on phosphors. These are only representative phosphors, and the full list contains many others with different characteristics. The P-22 is of prime interest because it is the one used for color CRTs, with all three versions included in order to achieve the color gamut. This leads the discussion to the subject of

TABLE 2.8 Characteristics of Registered Phosphors

JEDEC No.	Absolute Efficiency (Radiated Watt) Beam Watt × 100	Luminous Efficiency, k_b (Lumens) Beam Watt	Lumens per Radiated Watt
P-1	6	31.1	520
P-4	15	15	285
P-11	10–21	14–27	140
P-20	16	62.2	480
P-22 (G)	—	50	360
P-22 (B)	—	5	225
P-22 (R)	—	12	240
P-31	22	49.8	425
P-45	—	7.7	289

Sources: Matsumoto [13a], Karawada and Ohshima [13b], and Cola et al. [14].

color CRTs, which have a number of special characteristics described in the next section.

Color CRTs   Two types of color CRTs that have achieved any significant success: the shadow-mask and Trinitron versions. The first is the one invented by Law at RCA that took over the entire market for many years, until Sony engineers developed a different unit based on the one originally invented by Lawrence of Cyclotron fame, and to which Sony gave the name by which it has been known since. The basic structure of the original shadow-mask type is depicted schematically in Figure 2.11. The significant difference from the monochromatic CRT is that the phosphors are placed in groups of three on the faceplate, and backed up by a plate containing an aperture behind each triad of phosphor dots, whereas the single-color version has a single phosphor spread over the entire faceplate with no backup plate. The backup plate is termed the "shadow mask" because the electron beam can only pass through the holes in the shadow mask to strike the phosphors

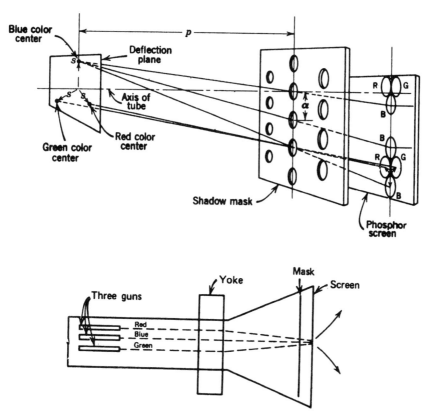

Figure 2.11   Shadow-mask color CRT. After Law [6], by permission of SID.

and therefore it acts as a mask whose shadow appears on the phosphor surface. The rest of the CRT differs only in that three separate electron beams are generated from three separate cathodes with only a single set of beam control, shaping, and acceleration electrodes for all three beams. Beam deflection is accomplished by means of a yoke placed as shown in Figure 2.10, and the same yoke is used for all three beams. It can be seen from the figure that each beam is constrained by the mask to strike only the phosphor that corresponds to the color represented by its beam and the three phosphor dots that make up each triad consist of three separate colors; red, green, and blue. Then the color gamut can be achieved by a proper combination of these three colors according to the equation

$$C = R_1(R) + G_1(G) - B_1(B) \qquad (2.12)$$

where C_1 = selected color
$R_1(R)$ = proportion of red
$G_1(G)$ = proportion of green
$B_1(B)$ = proportion of blue

In addition, three chromaticity coordinates are used that have the relationship

$$x + y + z = 1 \qquad (2.13)$$

The colors may then be plotted in terms of x and y as shown in Figure 2.12, which is known as the *CIE chromaticity diagram*, where CIE stands for the Commision International d'Eclairage, the international organization that establishes color display and other standards. The chromaticity coordinates in turn are related to the proportions of red, blue, and green by other equations; thus Figure 2.12 may be used to calculate the proportions. The locus of the colors has been established empirically, and the latest version, shown in Figure 2.13, differs somewhat from Figure 2.12 but is more accurate. Then, given the means for determining the proportions required for the different colors, it is possible to control the three beams so that the intensity of each corresponds to the required amount of each primary color to achieve the desired final color. It should be noted in passing that these primary colors apply to light-emitting sources and differ from those used in art that consists of light-reflecting surfaces where the primaries are red, yellow, and blue. However, the range of output colors is the same for both cases.

The Trinitron uses the same combinations of primary colors and three electron beams but differs from the shadow-mask in that the Trinitron has vertical slots for openings, the phosphors are laid down in vertical strips rather than dots, and an in-line gun is used. This arrangement is shown in

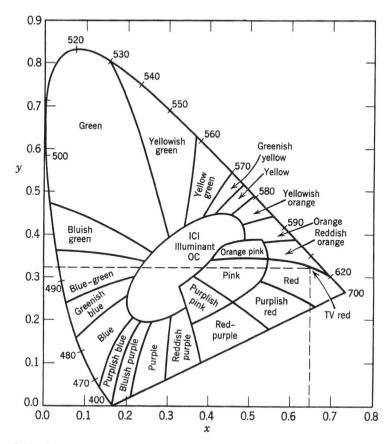

Figure 2.12 Location of colors in chromaticity diagram. After Pender and McIlwain [7], by permission of John Wiley & Sons, Inc.

Figure 2.14, where it is compared with the original delta gun structure. The Trinitron structure has been claimed to provide better resolution with an in-line gun, and this is substantiated to some extent in that the latest versions of the shadow mask use vertical apertures and phosphor strips, as shown in Figure 2.15, although the best overall resolution is still achieved with the delta gun and phosphor dots structure. In any event, the Trinitron has found its place as a display device for TV systems, although it has been restricted to Sony products until recently, when Sony decided to make the CRT available to other manufacturers of display monitors. As one result of this limitation, shadow-mask color CRTs have taken over the bulk of the market for TV systems.

In addition to these two main color CRT types, there have been a number of attempts to develop units that overcame some of the difficulties inherent

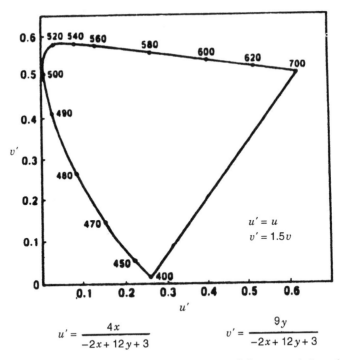

Figure 2.13   1976 CIE-UCS diagram. After Keller [8], by permission of SID.

in the manufacture of the units. Among the earliest was the *beam index* color CRT developed by Barnett and given the designation of "the apple tube." This unit employed only a single gun with a secondary simple beam to locate the phosphor energizing beam, and thus had promise to simplify the CRT structure and allow better resolution than either the shadow-mask or Trinitron structures. However, although units were manufactured of the early versions, and improved units have been made more recently, manufacturing difficulties and improvements in the standard types made this approach unsuccessful. The same can be said of another approach to a single-beam color CRT, namely, the *beam-penetration* type, where multilayer phosphors with different color outputs are energized by a single beam, to which different acceleration voltages are used to select the desired layer and color. However, limitations in the number of layers and therefore colors that could be selected, and the very high acceleration voltages that must be switched at high speeds, made this approach much too limited, especially as the high-resolution color CRTs to be used for HDTV and other applications have become available at reasonable costs. Therefore, these other approaches have been more or less forgotten as practical approaches, and the interested reader may consult the literature [4, 5, 9]. One other approach that has shown

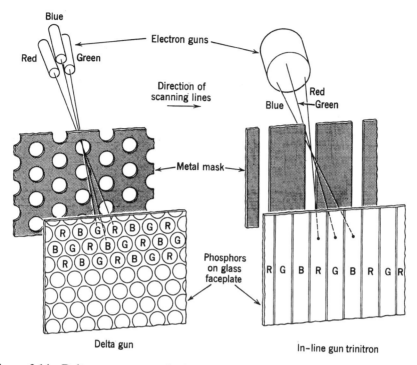

Figure 2.14 Delta gun compared with in-line gun Trinitron. After Herold [9], by permission of SID.

some promise is the one using a liquid crystal shutter to convert a CRT with a special multicolor phosphor into a unit that can accomplish much of what the standard units can in terms of the number of colors while using a single gun. However, any discussion of this technique is deferred to later sections concerned with FPDs. Therefore, we are left only with the two main approaches to color CRTs: the latest versions of the shadow-mask and Trinitron types with in-line guns. Attempts will surely continue to find other approaches, but for now these two techniques appear to satisfy the needs for direct-view color CRTs, with the latest versions capable of resolutions as high as 1280 × 1020 (pixels), at repetition rates as high as 70 Hz, and at sizes of up to 46 in. so that it remains for projection systems to achieve equivalent or better performance with larger screen sizes. These are discussed separately later in Section 3.5.2 so far as CRT systems are concerned, and in Section 3.5.3 where light-valve and LCD projection technologies are considered. Similarly, the ultimate result of CRT technologies, namely, the monitors in which the CRTs are used as the display device, are discussed in some detail in Section 3.3.2.

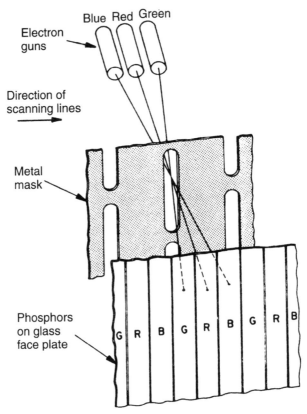

Figure 2.15 Vertical strip shadow mask. After Herold [9].

2.3.3 FPD Technologies

2.3.3.1 General Description

Flat-panel display technologies and displays exist in a much wider variety than CRTs units because of the various different technologies that are used for the FPD display panels and the many forms in which the display panels are found. A list of the most popular technologies is given in Table 2.1, parts *a* and *b*. At this point it is advisable to briefly describe the basic characteristics of the major FPD technologies as a preliminary to more detailed discussions of the specific FPD panels using the different technologies.

The first unique feature of FPDs is contained in the term *flat-panel*, denoting that the display unit is effectively flat in its depth dimension, especially as compared with the CRT monitor. It is true that some attempts have been made to produce CRT displays with the same form factor, but only with very limited success. These attempts continue, and some of the latest ones are described later, but at this point it is assumed that any display unit whose thickness is small compared to the other two dimensions is considered as a flat-panel display. It should be noted that there are examples of CRT FPDs, and the CRT technology is included in Table 2.1a. In addition, several unique features in CRT FPDs are not covered in Section 2.3.2.2. However, none of these CRT units has achieved much acceptance in the market, and they have been superseded by other technologies, so their importance to applications using FPDs is small. Therefore, any further treatment of CRT techniques as they apply to FPDs is discussed later in this section, along with the other less important technologies. This section concentrates on the five major FPD technologies that are listed in Table 2.1a. Field emission is a new technology with considerable promise, but it is still too untried to be considered a major technology and therefore is presented in Table 2.1b.

There are various basic configurations for FPDs, depending on what the technology is and the form of the product. Using the product type as the differentiator, the forms may be related to the A/N readout, A/N panel, and matrix panel as the three most important structures. Beginning with the first product type, its structure depends to some extent on what technology is used. For example, if LEDs are the display elements, they may be found in either monolithic or discrete forms, depending on the size of the character. Similarly, dc plasma has at least three different structures, and LCDs have two. The A/N panels are more restricted but may also vary, whereas the matrix panels have a wide range of structures. Therefore, it is more appropriate to consider the structures used for each product type separately in the sections devoted to the specific technology, and preface those discussions with a review of the pertinent operational principles applying to each major technology, beginning with the LEDs, and proceeding through EL, plasma, vacuum fluorescent, and liquid crystal displays. The lesser technologies are covered next in less detail, with the technical discussion followed by sections devoted to the product types using these technologies.

2.3.3.2 Light-Emitting Diodes (LEDs)

A full understanding of LED operation would require an in-depth discussion of semiconductor physics, which is inappropriate for this book. The interested reader is referred to a number of excellent texts on that subject [11, 13], and suffice it to state here that the LED is an outstanding example of the successful application of $P-N$-junction diodes to the production of electroluminescently generated light. A $P-N$-junction diode is one in which two layers of a solid-state material such as gallium arsenide (GaAs) are created with a surplus of *electrons* in one, and a deficiency known as *holes* in the other. These are termed charge carriers, and when the two layers are combined as shown in Figure 2.16, the resultant structure is passive because of the buildup of a reverse voltage across the junction due to the flow of charge carriers across the junction. If a voltage is applied in the positive direction of sufficient magnitude to overcome the reverse voltage, then current will flow as shown in Figure 2.17, which depicts the current–voltage characteristic of a

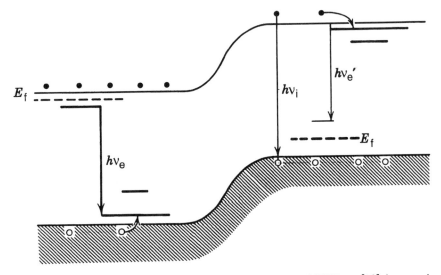

Figure 2.16   Forward-biased $P–N$ junction. After Piper and Williams [10], by permission of Academic Press.

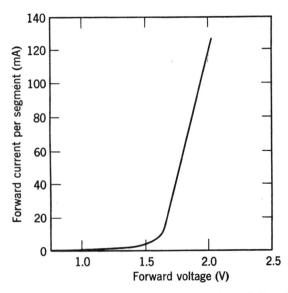

Figure 2.17   Current–voltage characteristics in forward-biased direction for a LED. After Goodman [11], by permission of SID.

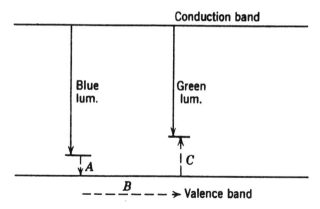

Figure 2.18 Energy-level diagram for a ZnS phosphor with blue and green luminescent centers. From Ivey [12], by permission of Academic Press.

LED. Under these conditions, the current injection causes the diode to emit in the spectrum ranging from the infrared to the ultraviolet, with the exact frequency depending on the material of which the diode is fabricated. The manner in which light emission is caused is illustrated in Figure 2.18, which shows two electron energy levels in the crystal, separated by what is termed the "forbidden band." When there is an excess of electrons in the E_2 level, radiactive recombination will occur to return it to thermal equilibrium. The frequency of the emitted light is given by the energy difference between the two levels divided by Planck's constant, and the effect of the different materials on the frequency is the result of different energy gaps. Table 2.9 lists several LED types manufactured from different materials, and the color and wavelength of each. It can be seen that all of the primary colors are attainable as well as some others, so that a full color range can be achieved. Of these, the blue LED is most recent, and its advent in conjunction with increased efficiency of all types has made multicolor LED panels at reasonable cost and light output with minimal power requirements possible, as is discussed further in the sections on specific product types.

TABLE 2.9   Visible LEDs

Color of Light	Wavelength (nm)	Material
Red	660	GaAlAs/GaAs
Orange	630	GaAsP:N/GaP
Yellow	590	GaP:NN/GaP
Green	565	GaP:N/GaP
Blue	470	SiC

This completes an admittedly minimal discussion of the physics of LEDs that only indicates the manner in which light is generated. However, it does show what the basic structure and material requirements are and what is available in terms of color types. From this information it is possible to proceed to a discussion of the product types and performance capabilities in the later sections, with recourse to the literature if further scientific detail is desired.

2.3.3.3 Electroluminescence (EL)

Electroluminescence as the means for producing light has been available for many years, and is well known in its night light and fluorescent bulb forms. It has also been used to create a variety of displays in its field excited form, but it was not until the development of thin–film electroluminescence (TFEL) that this technology achieved any large-scale applications and development of a number of different types of display formats. However, some review of the different forms of EL technologies is still of interest, and they are listed in Table 2.10. Of these, ac powder is the oldest, and is still of some interest to display systems as a means for generating backlights to be used with liquid crystal or other passive, non-light-emitting displays. This is the earliest form of electroluminescence, and is representative of the basic physical mechanisms involved. The basic structure, shown in Figure 2.19, consists of a luminescent layer made up of light-emitting phosphors dispersed in a trans-

TABLE 2.10   Electroluminescent Technologies

Ac powder
Dc powder
Ac film
Dc film

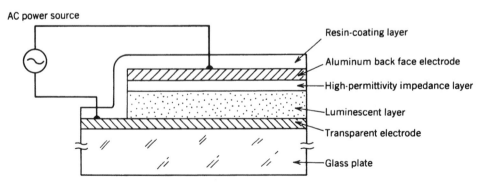

Figure 2.19   Structure of typical dispersed ac ELC. After Matsumoto [13a, p. 192], by permission of John Wiley & Sons, Ltd.

parent dialectric, and surrounded by transparent conductors that act as electrodes. When an ac voltage is placed across these electrodes, the phosphors will emit light in accordance with the empirical formula

$$L = K \exp\left(\frac{-C}{V^{1/2}}\right) \tag{2.14}$$

where L = luminance
V = voltage
K, C = constants

The exact wavelength of the light is determined by the particular material used as the EL source, and the wavelengths for several materials are shown in Figure 2.20. As noted, powder EL is used only as a light source, but the same phosphors are used for the other technologies so that the data given in Figure 2.20 hold for the devices based on those techniques as well.

The next EL technology shown in Table 2.10 has had a rather checkered career. Originally developed and sponsored by Dr. Aaron Vecht, it achieved some success in a variety of products built by Phosphor Products. However, it was not until it was taken over by Cherry Electrical Products that it resulted in truly viable products that warranted treatment as a serious contender for applications that may use EL devices for display. The structure is essentially the same as that for the ac version, differing primarily in that the electrodes are in direct contact with the EL layer and that the latter is activated by a dc

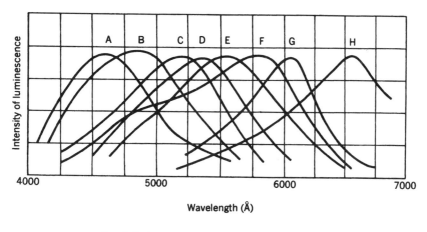

A — ZnS : Cu, Cl
C — ZnS : Pb, Cu (green)
E — ZnS : Mn, Cu (yellow)
G — Zn(S, Se) : Cu
B — ZnS : Pb, Cu (blue)
D — ZnS : Al, Cu
F — ZnS : Mn, Cu (orange)
H — ZnS : Cu

Figure 2.20 Distribution of luminescence wavelengths. After Matsumoto [13a, p. 192], by permission of John Wiley & Sons, Ltd.

rather than an ac voltage (see Figure 2.21*a*). In addition, it differs in that the layer undergoes a forming process when the dc voltage is first applied, resulting in the formation of a *P*-type copper sulfide layer so that a *P–N* diode is created at each copper particle. The equivalent circuit for the layer becomes as shown in Figure 2.21*b*, and a dc drive is sufficient, thus reducing circuit complexity to some degree. In addition, some claims have been made that lifetime and luminance output may exceed those available from the TFEL technology, discussed next. This is still somewhat contentious, but, in any event, Cherry has produced a line of products, now withdrawn, that appear to be competitive with the ac versions of the TFEL panels, as is discussed further in the later sections devoted to FPDs and monitors.

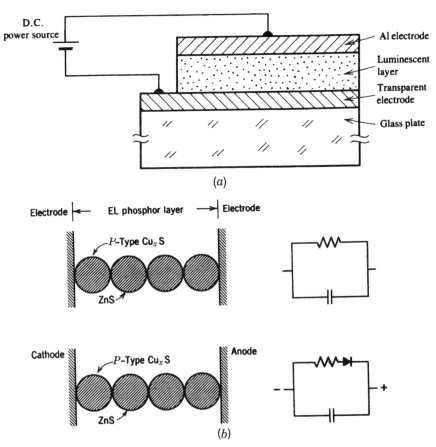

Figure 2.21 (*a*) Schematic representation of EL layer and its equivalent circuit before (top) and after (bottom) "forming process." After Karawada and Ohshima [13b] © 1973 IEEE by permission. (*b*) Structure of thin-film dc ELD. After Matsumoto [13a, p. 199], by permission of John Wiley & Sons, Ltd.

The ac TFEL technology is the third EL technology listed in Table 2.10, and differs from the first two in that the EL layer is laid down in the form of a thin film, evaporated onto an insulating layer that has previously been evaporated onto a transparent electrode, that in turn has first been evaporated onto a glass plate, as is shown in Figure 2.22. A second insulating layer is then evaporated onto the other surface of the EL layer, followed by a final evaporation of another transparent electrode on the insulating layer. This type of structure was first developed by Sigmatron in the 1970s, but with only limited success. The first fully successful ac TFEL unit is that initially developed by Sharp and followed by Planar and others. Both Sharp and Planar, which are the two major sources of TFEL products, now offer an extensive line of TFEL panels and monitor products that can be used for various applications in place of non-flat-panel CRT-based units. Panels with as large as an 18-in. diagonal and with full-color capabilities are becoming available, and are described more fully in the sections devoted to FPDs and monitors.

The last EL technology listed in Table 2.10 is that using dc TFEL. Its structure is quite similar to the ac types, and it was anticipated that it could achieve the performance of the latter along with the potential advantage of dc drive. However, it has suffered from thermal runaway and electrical breakdown, and it is included here mainly for information with the possibility that these problems may be solved at some time in the future.

2.3.3.4 Gas Discharge (Plasma)

Gas discharge displays have a long history, going back to the Nixie tube numerics of revered memory, and the ubiquitous neon signs that populate the highways and outdoors. However, the development of the more successful types, now termed *plasma display panels* (PDPs), are of more recent vintage. These exist both in the dc form as derived from the Nixie by

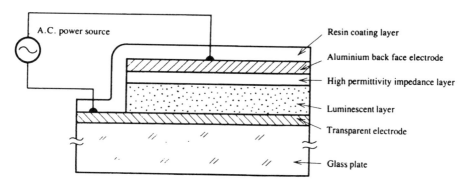

Figure 2.22 Structure of three-layer thin-film ELD. After Matsumoto [13a, p. 192], by permission of John Wiley & Sons, Ltd.

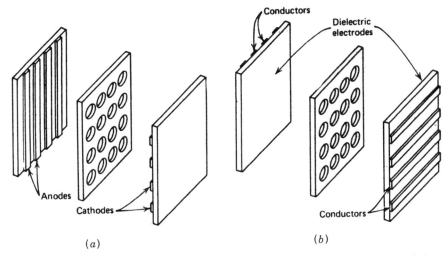

Figure 2.23 Gas discharge panel displays. After Cola [14], by permission of Academic Press.

Burroughs and Sperry, and in the ac versions, invented by Bitzer and Slottow, and given their initial impetus as a product by Owens Illinois. The basic structures of both the dc- and ac-driven types are shown in Figure 2.23a, b, and differ in these simplified forms mainly in that the dc version has the electrodes on the inside to permit direct connection to the gas, whereas the ac type has the electrodes on the outside on top of dielectric plates. Otherwise, they may be nearly identical, and the mode of operation also has many identical features and characteristics.

The physical basis for the operation of PDPs is that when a gaseous mixture of one or more rare gases, generally neon, is driven by a dc voltage across it to the breakdown point, it ionizes, freeing electrons that migrate toward the anode, with the accompanying positively charged ions moving to the cathode. The resultant condition of a tube containing the gas is demonstrated in Figure 2.24a, where the negative glow is the one that is seen. The breakdown voltage or ionizing potential is dependent on the gas used, and values for it and the first radiative excitation potential for several gases are given in Table 2.11, along with the characteristic color for each gas. The breakdown voltage (V_b) is given by

$$V_b = \frac{1}{\eta}\ln\left(\frac{1}{\gamma}\right) + V_i \tag{2.15}$$

where γ = first Townsend coefficient (number of electron–ion pairs per volt)

η = second Townsend coefficient (number of secondary electrons emitted from the cathode per bombarding positive ion).

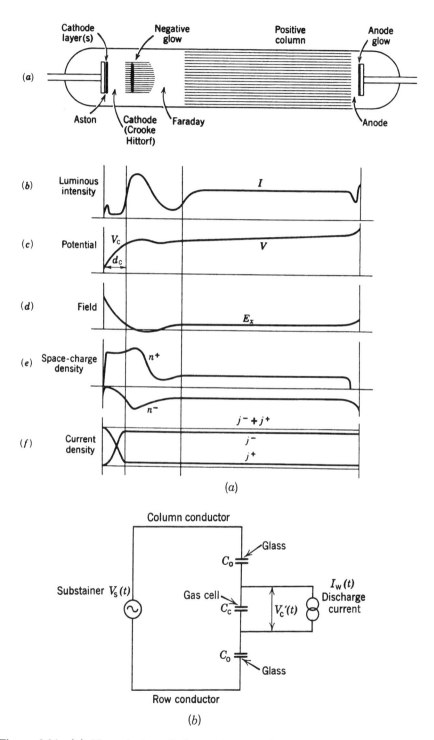

Figure 2.24 (a) Normal glow discharge in neon in 50-cm tube at $p = 1$ torr. Luminous regions are shown shaded. After Nasser [15], by permission of John Wiley & Sons, Inc. (b) Simple equivalent circuit of plasma cell. After Jackson and Johnson [16], by permission of Academic Press.

TABLE 2.11   **Characteristics of Gas Discharge**

Gas	Ionizing Potential (eV)	Excitation Potential (eV)	Color
A	15.7	11.6	Blue
Cd	8.96	3.78	Red
He	24.5	20.6	Yellow
Ne	21.5	16.7	Orange

Once broken down, that condition can be maintained by a lower voltage (V_s), termed the *sustaining voltage*, and a memory margin exists, defined by

$$\text{Memory margin} = \frac{V_b - V_s}{V_s} \tag{2.16}$$

The sustaining voltage is normally lower than the breakdown voltage by an amount that is the function of the gas and cell construction. The value of the memory margin is typically about 0.5, so that for a V_b of 150 V, V_s is 100 V.

As noted previously, the ac plasma cell differs physically from the dc type mainly in that the electrodes are on the outside of the glass plates, and the gas is separated from the inside of the plates by a dielectric layer. Under these conditions, the cell may be represented by the equivalent circuit shown in Figure 2.24b. The exact voltage across the cell capacitance (V_c) is given by

$$V_c = \frac{C_0}{C_0 + 2C_c} V_s = V_w \tag{2.17}$$

where V_c = sustaining voltage
$\quad C_0$ = glass capacitance
$\quad V_w$ = wall charge
$\quad C_c$ = cell capacitance

and the actual firing voltage (V_f') is expressed by

$$V_f' = V_f \frac{C_0}{C_0 + 2C_c} \tag{2.18}$$

where V_f is the external voltage. However, C_0 is usually much greater than C_c, so that the two voltages are essentially the same. Another equation of interest is the figure of merit, which is analogous to the memory margin (M)

for the dc cell, and is given by

$$M = \frac{V_f - V_c}{\frac{1}{2}V_f} \tag{2.19}$$

where V_c is the difference between V_f and the minimum sustaining voltage. A charge builds up on the wall that is positive during the first half-cycle, and is transferred to the other wall on the second half-cycle. Thus, the sum of the wall charge and the sustaining voltage is sufficient on each half-cycle to maintain the discharge and keep the cell in the light-emitting state. This type of display cell has proved to be quite effective, permitting large displays to be built as is discussed further in the sections on FPDs and monitors. However, it might be noted here that only one company has offered readouts using this technology, and it is essentially restricted to panel displays and monitors. Color has been achieved by adding multicolor fluorescent layers, and this technique is described later as well.

2.3.3.5   Vacuum-Fluorescent (VFDs) Displays

Vacuum-fluorescent displays (VFDs) might be considered as flat CRTs, but they differ sufficiently in structure and operation from the other flat CRTs to be considered as a separate type of FPD. They also go back many years, as did the plasma displays, and were originally single- and multidigit round tubes, in many ways similar to the Nixie. They were redesigned in 1974 by Ise and Futaba into flat units, and are now also found in both the segment and matrix addressed forms shown in Figure 2.25a and b, respectively. Referring to Figure 2.25a, this depicts the basic structure of the flat VFD, and consists of a cathode, grid, and anode, enclosed in an evacuated volume made up of the front glass and the rear substrate. This structure is very similar to CRTs in that the cathode emits electrons, creating an electron beam that is controlled by the grid, and which strikes the phosphor-coated anodes. Each anode segment is connected to a separate terminal, and can be made to emit light by applying a voltage to the desired segments. By this means each of the numerics can be created, using the seven-segment structure shown in Figure 2.25a or other characters using more segments. Larger arrays can be constructed as is described in more detail in the sections on readouts and panels. The structure of the dot-matrix version illustrated in Figure 2.25b contains a cathode, a diffusion grid, and anodes, but the anodes are selected by means of row electrodes and column control grids that enable each dot to be individually selected, using matrix addressing. The phosphors are coated on each of the anode's dots, and different colors can be achieved by using different phosphors. A number of the phosphors that are used are shown in Figure 2.26. One unique feature of VFDs that differentiates them from standard CRTs is that the anode voltage is much lower, in the range of 25 V as compared with the kilovoltage needed for the CRTs, making this technology convenient to use for FPDs. Therefore, they are included in this section.

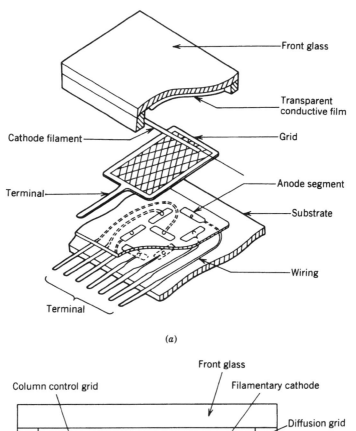

(*a*)

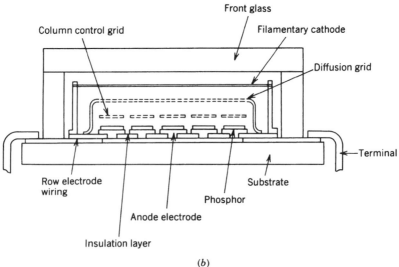

(*b*)

Figure 2.25 (*a*) Basic structure of flat VFD. (*b*) Normal-type dot-matrix VFD. After Morimoto and Pykosz [17], by permission of SID.

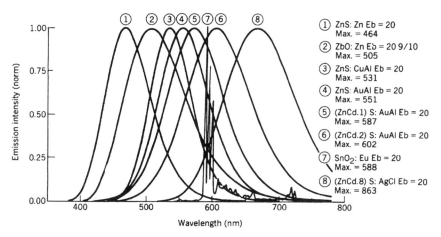

Figure 2.26 Emission spectra of some phosphors. After Morimoto and Pykosz [17].

2.3.3.6 Liquid Crystal (LC)

Liquid crystal displays (LCDs) are available in a variety of forms, using a number of different aspects of the liquid crystal technologies. First, it should be emphasized that all LCDs are passive in that they do not emit light, but rather either block or transmit light coming from some other source. This source may be part of the display, or the light may be obtained from ambient illumination. Although the liquid crystal phase was discovered as long ago as 1888, it was not until the early 1970s that Heilmeier and his associates at RCA Laboratories (now Sarnoff Laboratories) discovered ways to use this material for display devices. The earliest technology was that termed *dynamic scattering*, a nematic type in which the molecules were initially aligned in either the homeotropic or homogeneous arrangement and went from transparent to milky white when activated. This technique was used for a number of devices, but was found inadequate for most potential applications and began to be discarded. It is now completely gone, and this disappearance from the industry might have meant the end of LC technology. However, this approach was followed by the invention of the twisted nematic LCD, for which a number of investigators [19–21] have claimed credit. Further developments have resulted in supertwist and ferroelectric versions, with the net effect that LCDs have become the most prevalent types of FPDs, in particular for those applications that require battery-driven, low-power operation, because a major characteristic of LC display products is that they require only microwatts of power if no external light source beyond that provided by ambient illumination is required. Color versions have been developed using assemblies of light filters, and drive circuits termed *active* and *passive matrix* have led to a number of display products used in particular in portable computers, but also found in a variety of instruments and other handheld

Some types of fluid crystals

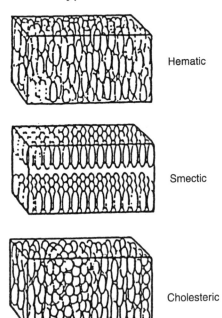

Hematic

Smectic

Cholesteric

Figure 2.27 Schematic representation of ordering in the three mesomorphic states. After Schlam [18], by permission of SID.

equipment. Thus, the liquid crystal technology is of great importance for display applications, and warrants considerable attention, within the limits of space in this text.

Briefly, then, liquid crystals look and pour like ordinary liquids, but they are complex organic chemicals and differ from most liquids in that they assume one of the three crystal orientations shown in Figure 2.27. The basic structure of an LCD cell is shown in Figure 2.28, and this was the form in which the initial displays were made, before the invention of the twisted nematic (TN) type. However, the initial structure was subject to various defects, and these were overcome by going to the structure shown in Figure 2.29, where the addition of the polarizer–analyzer combination created a viable cell. The mode of operation is illustrated in Figure 2.29, where the molecular axis of the LC material is made to rotate continuously through 90° from one plate to the other. The input light is linearly polarized by the input polarizer so that with the LC molecules in the passive state shown in Figure 2.29*a* the polarized light will be rotated by the LC cell through 90° and the light will be blocked by the analyzer if it is parallel to the polarizer. It might be noted here that a *polarizer* is an optical element that passes light of one polarity and blocks light of another polarity, so that nonpolarized light will be polarized by passing through the polarizer. Then, when a voltage is applied across the cell, the LC molecules will be untwisted, as shown in Figure 2.29*b*,

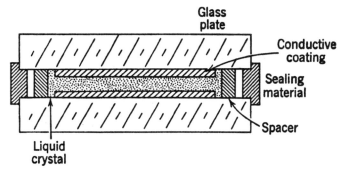

Figure 2.28 Schematic cross section of an LCD. After Goodman [19a], by permission of SID.

and the plane of polarization of the polarized light will not be twisted. Therefore, it will pass through the analyzer and can be seen, creating a light on dark display. The opposite can be achieved by placing the analyzer at a right angle to the polarizer. In both cases, there is considerable attenuation of the light due to losses in the polarizer–analyzer combination and the viewing angle is quite limited. However, this was found to be at least minimally satisfactory, and a number of FPDs using this structure have been included in a variety of products, in particular portable, battery-operated instruments.

There were some additional problems in the initial TN structure besides the light loss, particularly in attempting to use matrix addressing techniques.

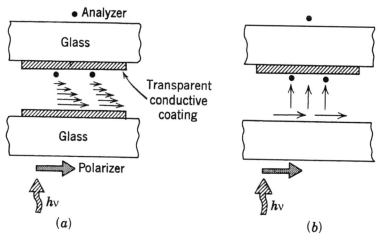

Figure 2.29 Side views of twisted nematic for (a) $V = 0$ and (b) $V > V_{TH}$. After Goodman [19b], by permission of SID.

Unfortunately, the structure lacked clearly defined breakpoints so that significant crosstalk could occur. The next step was to use twists beyond 90° extending the twist to as much as 270°, and creating the type of molecular alignment shown in Figure 2.30. This type of cell has been termed the *supertwisted* and *double-supertwisted nematic* or *supertwisted birefringence effect* (STN or SBE). The result has been some significant improvement, and cells using these effects have superseded the simpler TN versions. The structures for the double-supertwist and monochrome supertwist cells are shown in Figure 2.31. These cells have the same basic form as the TN versions, but the polarizers must be aligned in an off-axis polarization to achieve improved contrast ratios. The use of 30° for the front polarizer and 60° for the back polarizer results in a cell with a yellow color in the nonselect state and black in the select state. In addition, rotating the second polarizer by 90° results in a purplish blue output color for the nonselect state and colorless for the select state. The color compensating plate shown in the double-supertwisted version compensates for these color effects, and the result is a white-on-black display with a significant improvement in contrast ratio and viewing angle. Therefore, this is the preferred arrangement in terms of performance but leads to a more expensive cell. The solution is the supertwist display with an optical retarder that retains the white-on-black display and the other improvements at a low cost.

The next type of LC display that has achieved acceptance is the ferroelectric or smectic display. This technique is more recent than the previous two, but has shown considerable promise and may displace the others as they have

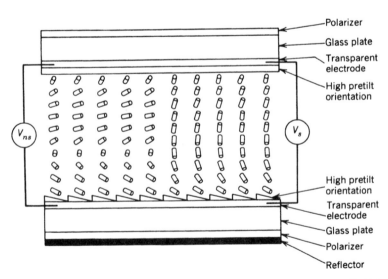

Figure 2.30 Schematic view of a supertwist STN display. After B. S. Scheuble, "Liquid Crystal Displays with High Information Content," *SID Sem. Lect. Notes*, 1991, F-2-25. By permission of SID.

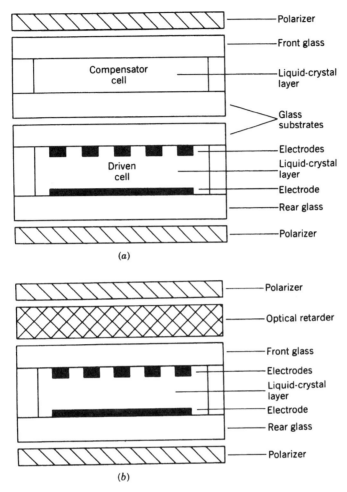

Figure 2.31 Double-supertwist (*a*) and monochrome supertwist (*b*) construction. After Pryce [20]. Reprinted from *EDN Magazine* (Oct. 12, 1989). Copyright 1997, Cahners Publishing Co. A division of Reed Elsevier, Inc.

displaced the initial TN, which in turn displaced the earliest dynamic scattering versions. In this technology, a thin layer of chiral smectic LC molecules termed the *surface-stabilized ferroelectric* (SSF) is used. Its main characteristic is that the average orientation of the molecules can be switched between two states, as illustrated in Figure 2.32. The light transmission is expressed by

$$I_0 = I_t \sin^2 4\theta_0 \sin^2 \pi\Delta\frac{nd}{\lambda} \tag{2.20}$$

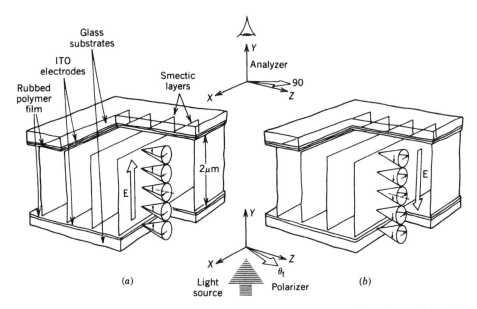

Figure 2.32 The two uniform parallel states characteristic of surface-stabilized devices formed by chiral tilted smectic mesaphases: (*a*) uniform "up" state; (*b*) uniform "down" state. After Crossland et al. [21], by permission of SID.

where θ = cone angle of the chiral smectic
     λ = wavelength of light

One desirable characteristic of this LC technology is that it can attain relatively fast switching speeds, in the range of 10–100 μs, which exceeds that possible with any of the TN types. As a result, it offers great promise of becoming a preferred technology.

Another application for smectic LCD technology is the thermally addressed light valve. However, as this falls into the area of projection displays, it is discussed in further detail in Section 3.5, on large-screen systems. However, at this point it should be noted that the technology employs the characteristics of smectic LCDs that have to do with what happens to a cell consisting of a thin film of smectic material when it is heated into the isotropic state and then cooled. Under these conditions, it will become either cloudy or transparent depending on whether the cooling is rapid or gradual. Under these conditions, the cell may be used as a light gate and incorporated into a projection system. This type of projection display system usually referred to as a *light-valve*, has found fairly extensive use in large-screen systems, in general based on one of the TN technologies, in particular the supertwisted one, where the projection element is basically a standard LCD, so designed as to be usable with an optical system, as described further in Chapter 3.

2.3.3.7 Other LCD Technologies

Several other LC technologies have been used to develop electronic displays although with limited success. However, they do deserve at least some minimal mention as they were considered promising and do incorporate some interesting capabilities. First among these is the one termed the guest–host (GH) display. Its major distinction is that it offers a way to achieve color without having to include complex drive circuitry and color filters as are required by the active-matrix display described later. The basic principle of the GH cell, illustrated in Figure 2.33, is the inclusion of dye molecules that align themselves parallel to the LC molecules. Then, the dye molecules will be reoriented when the LC molecule is realigned by the application of a voltage, and the cell may be made to transmit either colored or white light. Under these conditions a combination of three cells with different-colored dyes might be used to create full-color displays. However, it never was brought to this point because the active-matrix approach has been found to be more successful.

The last LC technology considered here, termed *nematic curvilinear aligned phase* (NCAP), is the only one still using the dynamic scattering phenomenon. It differs from the previous versions in that very small spheres of nematic liquid are encapsulated in a polymer matrix, and then laminated between large panels of a flexible plastic film. Then, each capsule acts as a light-switching element, with the light scattered in the nonenergized condition, and transmitted without change in the energized state. This is quite similar to the standard, nontwisted cell, but differs in that interactions between the polymer and the liquid crystal in the nonenergized state create the alignment. In addition, large assemblies may be built because of the flexible structure, and faster switching speeds as well as operation at lower temperatures than the standard cell are possible. A number of products have been developed, but in general this technology has not captured a significant share of the market.

2.3.3.8 Field Emission (FE)

A new approach to FPDs is represented by a number of developments that use field emission as the basic technology. This is somewhat similar to the

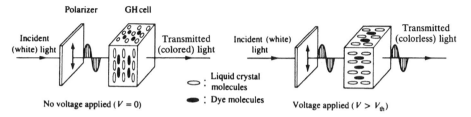

Figure 2.33 Principle of guest–host (GH) display. After Matsumoto [13a, p. 43]. Reprinted by permission of John Wiley & Sons, Ltd.

CRT technology in that one or more electron-emitting cathodes are involved, but differs in that the CRT cathode used thermionic emission, whereas the field emission technology may use cold cathodes made of molybdenum or similar material, with the structure organized as shown in Figure 2.34, which demonstrates another similarity between FEDs and CRTs, in that the light output comes from the phosphor faceplate that is struck by the electron beams. Finally, there is also the equivalent of the CRT control grid between the emitters and the faceplate in what may be termed a *gate* or *extraction grid* that controls the intensity of the electron beam. This type of FED [22–24], similar to earlier versions developed by Northrup and Matsushita is much more effective in structure and operation.

In addition to the FEDs using cold cathodes as the emitter, there are several other types that use other materials. Prime among them is the one that uses diamond as the emissive material, organized as shown in Figure 2.35. An amorphic diamond thin film is laid down on a network of patches, as shown in Figure 2.36, and the electron emission from the patches forms a beam that impacts on the phosphor and causes it to emit its characteristic color. The diamond is considered to have physical characteristics that render it superior to other materials in that it requires less current and can be operated at higher temperatures than most of other materials appropriate for FEDs. These are very promising developments, and although these devices are still rather far from being fully marketable, the basic technology appears to be quite feasible.

2.3.3.9 Other Technologies

Several other technologies have been used to produce FPDs with varying amounts of success. Although none of them have reached the point of wide acceptance as shown for the five major technologies, in particular LC in its various manifestations, they have led to devices with some desirable characteristics and warrant brief discussion in this review. They are all of the nonemitting type and operate by either transmitting or reflecting light, depending on the state of the cell. In addition, they all follow the matrix assembly structure and drive in the full display, and use similar physical arrangements to contain the active material. The major types are the electrochromic (EC), electrophoretic (EP), suspended particle (SP), and electromagnetic (EM), with the last one differing from the others in that it involves the physical movement of a discrete element. To these might be added incandescence (IN), which is still being used, albeit to a very limited extent in signs, message displays, and scoreboards.

Beginning with EC, the technology operates by using the phenomenon of change in light-absorption characteristics of certain materials when subject to an applied field. These materials are transparent in the nonactivated condition, and absorb some color when a field is applied. The color remains when the field is removed, but may be made to revert to the transparent condition by the application of a field with opposite polarity. Another possibility is for

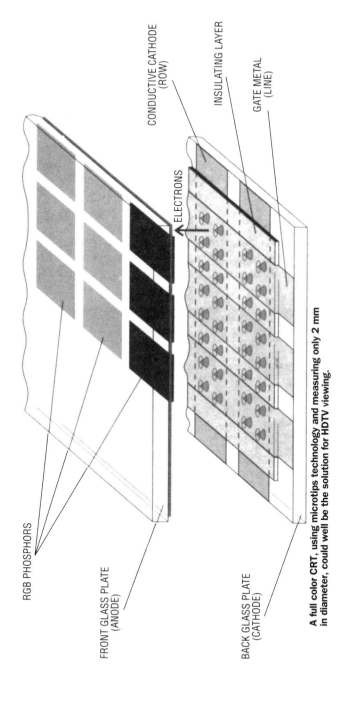

CONDUCTIVE CATHODE
(ROW)

INSULATING LAYER

GATE METAL
(LINE)

ELECTRONS

RGB PHOSPHORS

FRONT GLASS PLATE
(ANODE)

BACK GLASS PLATE
(CATHODE)

A full color CRT, using microtips technology and measuring only 2 mm in diameter, could well be the solution for HDTV viewing.

Figure 2.34 Full-color CRT using microtips technology. After Mannion [22], by permission of IEEE.

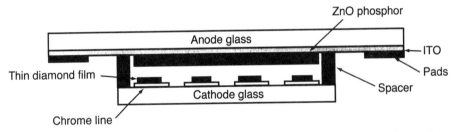

Figure 2.35 Basic structure of the diode device based on diamond thin films. After Kumar et al. [23], by permission of SID.

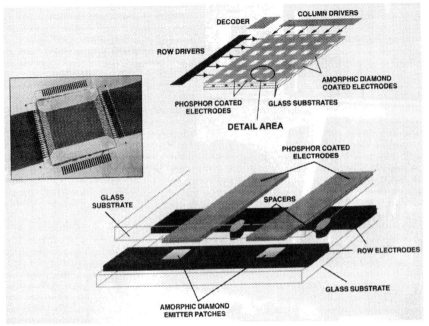

Figure 2.36 Detail of the structure of flat-plate display developed using diamond technology. After Chan [24]. Copyright, courtesy of Intertec Publishing, Overland Park, KS 66212. All rights reserved by Curtis Chan.

the material to change color in the two conditions of operation. Thus, this is a display device that can be operated to create a variety of images. The basic cell structure for this type of device is shown in Figure 2.37, and a variety of organic and inorganic materials are available for the electrochromic film. Several devices based on these principles have been developed and exhibit attractive appearances. However, the lack of a well-defined threshold makes them difficult to use in matrix assemblies, and the response times are too

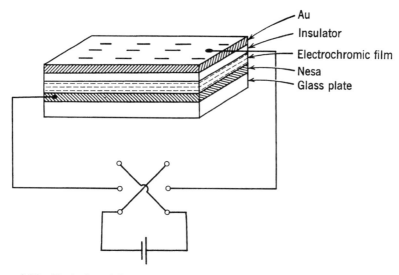

Figure 2.37 Typical multilayer solid-state electrochromic structure. After Deb [25] © 1974 IEEE.

slow for many applications. Therefore, they remain an interesting but limited group, and the technology has languished.

Next is the EP technology, which is similar in operational characteristics to EC, although the visual results are obtained by different means. The phenomenon in this case is electrophoresis or the motion of charged particles in a fluid under the influence or an electric field. One structure that has been used is shown in Figure 2.38, and the similarity to that use for EC is apparent. However, the light-affecting material consists of some type of colored pigment particles contained in a suspending medium. When a field is applied, the charged pigment particles will migrate to the transparent electrode, changing the color of the cell. Again the image can be quite attractive, but limitations in speed, number of colors, and life have severely restricted the usefulness of displays based on this technology. However, it remains as an approach that may become more feasible in the future, and is worth at least this limited discussion.

The next technology in this group is the SP, which has definite similarities to both of the previously discussed approaches, but appears to have some better possibilities for meeting certain application needs. The term *colloidal display* is also applied to this technology, and the operating principle is the alignment of colloidal particles parallel to an ac field when it is applied across a cell containing the particles. Under this condition the cell is transparent to light, whereas in the absence of a field, the particles are randomly oriented and the light is absorbed. This is somewhat similar to the guest–host (GH) LCD technology, but it is claimed that the cell is simpler

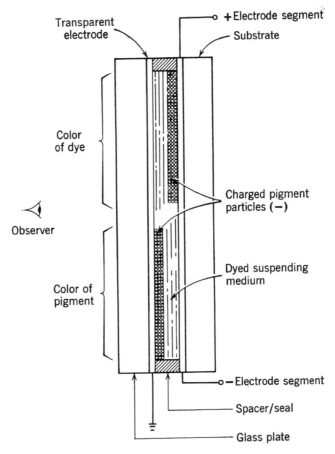

Cross section of an
electrophoretic image display cell

Figure 2.38 Schematic of simple EPID cell. After Dalisa [26] © 1977 IEEE.

and cheaper to manufacture, can be produced in larger sizes, and is more reliable than the other. The cell structure is similar to those of EC and EP, consisting of a transparent front electrode, a rear electrode, and glass walls within which the particles are suspended in a liquid. However, there is too little experience with this technology to draw conclusions with assurance, and it remains an interesting but unproved technology.

Finally, in this group of passive display technologies there is the EM technology, which differs considerably from those of EC, EP, and SP in its basic aspects, although it is similar in that light reflection is used for creating the observed image. However, in this case the controllable element is a disk

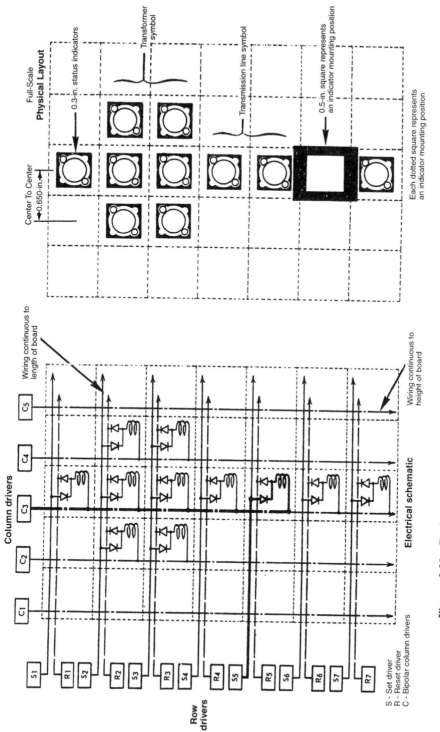

Figure 2.39 Basic structure of electromagnetic matrix element. Courtesy of Ferranti-Packard.

or other physical surface that can be rotated so as to present one of two possible positions to the light source. Thus, by using different colors on the two surfaces, it is feasible to create recognizable shapes from matrix assemblies, or select one of two possible shapes contained on the surfaces. The basic structure of an element containing a rotating disk is shown in Figure 2.39, in which the element is rotated by means of applying a current pulse to a coil that creates a magnetic field that then orients the disk by means of the fixed magnet attached to the disk. This type of assembly has seen extensive use in message displays, both indoors and outdoors, where its ability to be combined into large, matrix-addressed displays has made this technology quite effective for such applications. Its major drawbacks are the rather large currents required to activate numbers of disks, the slow response, and the relatively high cost of a total system. However, the technology continues to be used in various forms because of its ruggedness and good legibility.

The last of the technologies, referred to above is IN, which is probably the oldest and is being supplanted in most applications for which it is appropriate by other technologies, especially LEDs. However, there are still applications in which the high luminance capabilities of standard lightbulbs makes it possible to create large digit displays in outdoor environments that can be read more easily than other technologies that are capable of producing large-board displays. One example is scoreboards found in sports arenas, and another is animated signs. Thus, it remains both a feasible and useful technology, albeit quite limited in application. The basic technology is that used for ordinary lightbulbs, consisting of filaments in a gas-filled bulb, and the usage for electronic displays differs only in that the voltage used may be lowered to achieve longer life. Some applications (e.g., aircraft panels) also use small specialized bulbs, but this type of application is becoming rather uncommon.

2.3.3.10   Conclusions

We have examined five major, one potentially major, and four minor FPD technologies, and it is of some value in concluding this review to present a table of comparisons of some of the performance characteristics exhibited by each of these technologies. These are expressed in terms of a set of parameters that specify the relevant capabilities in measurable terms. A useful addition would be legibility, but this is not readily available for each technology alone, as this parameter requires full panels for meaningful results. However, contrast ratio is one possible measure, and is included as the closest instrumental parameter that can be used to establish comparative legibility. There may also be other parameters of interest, but the ones chosen here should be sufficient to allow meaningful comparisons of the performance capabilities of the different FPD technologies and provide a means for determining which technologies are at least adequate for the specific applications. The actual technology evaluations in terms of the

TABLE 2.12 Comparison of FPD Technologies

Parameter	LED	EL		Plasma		VF	IN
		Ac	Dc	Ac	Dc		
Voltage (V)	1–5	150–250	150–250	90–150	180–250	12–40 dc	10–120 ac
Current	1–10 mA	1–10 mA	1–10 mA	1–10 mA	1–10 mA	2–5 mA	0.1–10 A
Contrast ratio	30–50	20–50	15–20	15–30	10–20	30–50	5–15
Response time	$<1\ \mu s$	$10–40\ \mu s$	$10–50\ \mu s$	$10–20\ \mu s$	$10–20\ \mu s$	$10–15\ \mu s$	100ns–1 s
Luminance (nits)	30–300	25–250	100	150	150	150–1000	100–1000
Colors (no.)	4	4–16	4–16	1	1–4	5	1

Parameter	LC	EC	EP	SP	EM
Voltage (V)	2–5 ac	0.5–3 dc	−50 dc	5–10 ac	1–5 dc
Current	$1–10\ \mu A$	$5–10\ mC/cm^2$	$10\mu A$	$1–10\ \mu A$	> 1 A peak
Contrast ratio	10–20	10–20	15–25	10–30	10–25
Response time	25–250 ms	500 ms	50–150 ms	90–300 ms	100ms–5 s
Luminance (nits)	—	—	—	—	—
Colors (no.)	All	3	3	2	3

TABLE 2.13   Comparison of Performance Capabilities of FPD Technologies

Capability	LED	EL	Plasma	VF	IN	LC	EC	EP	SP	EM
Full color	No	Yes	No	Yes	No	Yes	No	No	No	No
Large area	Yes	No	Yes	No	Yes	No	No	Yes	Yes	Yes
Large capacity	Yes	Yes	Yes	Yes	No	Yes	No	No	No	Yes
High resolution	No	Yes	Yes	Yes	No	Yes	No	No	No	No
Display quality	Fair	Good	Good	Good	Fair	Good	Fair	Good	Fair	Fair

specific applications are discussed in the chapters that deal with the respective applications, and the comparisons shown in Table 2.12 are intended primarily to be introductory.

These data provide some clues as to which technology might be most appropriate for an application, and also provide a basis for developing another table that compares the technologies with respect to generalized performance capabilities and characteristics given in terms of language descriptions. These are shown in Table 2.13.

It can be seen from these evaluations that EL, plasma, VF, and LC appear to be best for many applications, but some of the others have special capabilities such as speed and high luminance for LEDs, low power for EC, and the ability to create large displays for LEDs, IN, SP, and EM. Therefore, the data contained in Tables 2.12 and 2.13 alone are not sufficient to allow one to make a fully educated choice but they are good starting points, supplemented by additional data found in the chapters dealing with specific applications.

2.4   MULTIPLEXING AND MATRIX ADDRESSING

2.4.1   Introduction

Although multiplexing and matrix addressing are, strictly speaking, not FPD technology terms in the same sense as those covered previously, they are so important to the proper operation of all types of FPD readouts, panels, and monitors that they warrant separate treatment in this technology section. Specifically, the basic considerations behind these modes of operation are of importance here, although some special characteristics and requirements may exist for specific product applications. These are discussed in greater detail in the sections dealing with the specific product types found in Part 2. The discussions in the following sections are concerned with the basic configurations required for multiplexing and matrix addressing, as well as the advantages and limitations of these modes of driving flat panels of any type or using any of the specific FPD technologies covered previously. In certain respects this has as its equivalent the circuitry used to drive CRTs when used in monitors, and it might be appropriate to discuss those techniques in this chapter. However, that circuitry is more specific to the particular type of CRT used in the monitors, and does not have the same generality as exists for multiplexing and matrix addressing, so it seems more appropriate to defer the CRT discussion to Section 3.3 on monitors.

2.4.2 Multiplexing

Multiplexing refers to a technique for reducing the number of drivers needed to energize a multiplicity of display elements by sharing them among the elements on a time-division basis and may be defined as any repetitive scanning sequence in which the display elements are sequentially addressed, either one element at a time or a group of elements at a time. In some respects this technique is analogous to the scanning sequence used in CRT displays, where the electron beam corresponds to the drive and the sequential addressing is achieved by means of the scanning beam. Similarly, when multiple beams are used, they correspond to the group of elements. However, the CRT version is limited to devices with electron beams, whereas multiplexing in general may be applied to a variety of electronic components, such as solid-state memories, for example.

The multiplexing procedure illustrated in Figure 2.40 contains an n-digit, seven-element assembly of which one digit is shown. The backplane has one line for each digit, which is used to switch to the digits in sequence using a digit multiplexing signal with the appropriate selection of the digit elements for each of the eight positions as the digit select signal multiplexes through the sequence. The advantage of this approach is in the reduction of the number of leads required to drive the assembly. For example, if each of the

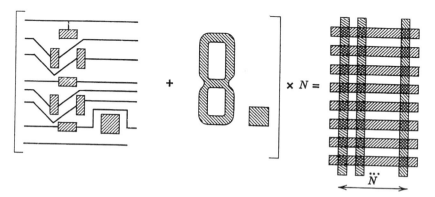

Figure 2.40 Example of interconnection pattern for front and rear electrodes of multidigit, multiplexed seven-segment numeric display. After Kmetz [27], by permission of Plenum Press.

eight digits were connected separately, this would require $8 \times 7 = 56$ leads, whereas the multiplexed scheme allows the same result with only $7 + 8 = 15$ leads. However, there is a loss in contrast ratio and a limit in the number of digits that can be multiplexed. The first is given to a good approximation by the duty factor times the nonmultiplexed contrast ratio, or

$$C_r = C_m \frac{1}{n} \tag{2.21}$$

where C_r = contrast ratio for multiplexed digits
  C_m = contrast ratio for nonmultiplexed digit
  n = number of multiplexed digits

This is accurately valid for fast response characters such as those using LED technology, but may be somewhat worse for slower-response units such as those using LC technology, due to the retention time maintaining the voltage level during the off time. However, even at best it is apparent from Equation 2.21 that the multiplexed contrast ratio for an eight-digit unit is only 12.5% of the maximum, so the peak luminance must be increased by a factor of 8 to achieve the same contrast ratio. This limits the number of digits that may be multiplexed while still maintaining adequate contrast ratios, especially for LCD units where the relevant equation is

$$\frac{V_{on}}{V_{off}} = \left(\frac{8}{N} + 1 \right)^{1/2} \tag{2.22}$$

where N is the number of digits and the maximum voltage ratio is 3, using a multiplexing technique termed $3:1$ *matrix addressing*, which is described in more detail in the section devoted to matrix addressing. Therefore, the contrast ratio for an eight-digit LCD will be only 47% of the maximum contrast ratio attainable for one digit. As a result of this situation, whereas the 7- or 16-segment light-emitting displays such as dc plasma and VFD character assemblies can be obtained in quantities as high as 32 multiplexed characters, the LCD units are generally limited to a maximum of eight to keep the contrast ratio close to 10. In general, however, the higher quantity character displays are handled by dot-matrix techniques.

2.4.3   Matrix Addressing

2.4.3.1   *Basic Techniques*
A *matrix* is an assembly of elements arranged in some geometric pattern, such as the basic $X–Y$ square array shown in Figure 2.41. This arrangement can be used to drive two terminals of any type of device and is widely used to select memory elements in solid-state memories. The connections are made to two terminals of the element, and if the proper signal is directed across

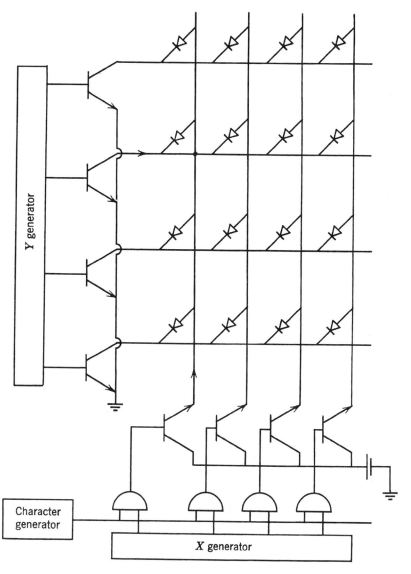

Figure 2.41 Basic *X–Y* matrix configuration. After S. Sherr, *Electronic Displays*, 2nd ed., 1993, p. 278. By permission of John Wiley & Sons, Inc.

any element, that element is selected and will change its state. In the case of the various types of FPDs covered in this and other chapters, the connection points consist of the electrodes connected to the two sides of the display, arranged in this matrix fashion. Thus, when the *X* electrode and *Y* electrode facing each other are driven by the appropriate signal, the volume between the two electrodes is energized and will emit light for the light-emitting types, and either transmit or block light for the passive types. Thus, it is possible to select a group of elements arranged in an $X-Y$ square array by means of $X + Y$ connections rather than $X \times Y$ connections as would be required in a nonmatrix-addressed array. Therefore, a 1000×1000 array would require only 2000 connections for matrix addressing rather than the 10^6 connections needed for the nonmatrix-addressed arrangement. This is of no importance for color CRTs that require only three connections, but is of great importance for FPDs, including the matrix-addressed CRTs. The result is that all FPDs are matrix-addressed in some fashion, and, as previously noted, multiplexing is the simplest type of matrix addressing, combined with scanning; the latter technique may also be used with larger arrays to further simplify the drive circuitry. One problem that arises when matrix addressing is used is the need for well-defined breakpoints between nonactivated and activated drive levels. Unfortunately at least one of the most popular technologies, that is, liquid crystal, does not have very well defined breakpoints, as is illustrated by Figure 2.42 for the general case. This figure illustrates the optical response of ideal and practical display elements to the application of a drive signal. Thus, it can be seen that if the drive signal on the row line is below V_{on} with the column line at zero, the display element will not be turned "on." However, if both the X and Y lines are driven by opposite-polarity signals so that the sum of the two signals exceeds the required "on" drive voltage, the element will be activated and either emit light or act as a light gate, depending on whether an active or passive technology is used. As noted previously, the commonly used active technologies exhibit transitions that are sufficiently sharp to permit clear differentiation between the two conditions by using the criteria

$$V_{off} < V_{on}, \qquad 2V_{off} = V_{on} \text{ (saturation)} \tag{2.23}$$

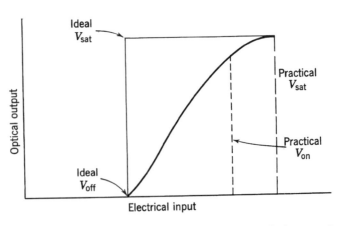

Figure 2.42 Comparison of ideal and practical responses of electrooptical device. After S. Sherr, *Electronic Displays*, 2nd ed., 1993, p. 279. By permission of John Wiley & Sons, Inc.

These conditions apply reasonably well to the active technologies, but for the passive technologies, in particular LC, the converse is true, and it is necessary to use various expedients in order to arrive at reliable matrix addressing. This has led to the use of separate matrix elements connected to the X and Y conductors on the two sides of the FPD. However, the other popular technologies do not require such additional drive elements, and further description of what is termed *active-matrix addressing* is deferred to later in this section when the special requirements of matrix panel LCDs are discussed. At this point the descriptions are restricted to the general approach that applies to the other technologies.

Returning to Figure 2.41, if the drive signals applied to an X and a Y line meet the criteria established by Equation 2.23, the diode connected between the selected lines will be driven so that if it is a LED, it will emit light. Similarly, if an EL panel or a gas takes the place of the diode array, the point between the two lines will emit light. Then, drive groups of sequential lines can be used to generate an image which can be retained if the technology has internal memory, or refreshed at a proper rate to avoid flicker. However, unless all the drive lines are activated in parallel, it is necessary to consider the duty factor effect on contrast ratio for the same reasons that apply to multiplexing. In order to minimize this effect for large arrays, it is possible to drive the FPD matrix a row or a column at a time with the other set scanned a pixel at a time, resulting in a TV-type display. Other sequences are also possible to minimize the contrast ratio effect, such as driving all columns in parallel while scanning the rows, but at the cost of additional column drivers. Finally, insofar as contrast ratio is concerned, there is another aspect of matrix addressing known as the *discrimination ratio* which is the ratio of the "on" to the "off" luminance, given by

$$\text{DR} = \frac{L_{\text{on}}}{L_{\text{off}}} \tag{2.24}$$

DR is then a measure of how well an "on" pixel can be differentiated from an "off" pixel when the matrix array is addressed on a row-at-a-time basis. The duty factor for this approach is given by $1/n$ for a matrix with n rows and m columns, and the contrast ratio is

$$C_S = \frac{L_{\text{on}}}{nL_{\text{off}}} + \frac{n-1}{n} \tag{2.25}$$

or, if D is substituted into Equation 2.25 then

$$C_S = \frac{D}{n} + \frac{n-1}{n} \tag{2.26}$$

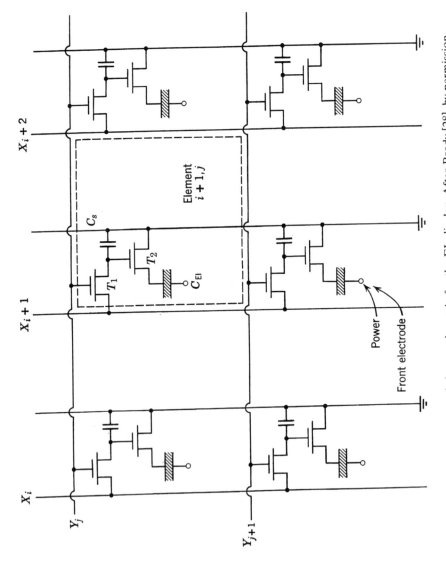

Figure 2.43 Design of elemental matrix for the EL display. After Brody [28], by permission of SID.

and, if n is much larger than 1, then

$$C_S = \frac{D}{n} + 1 \qquad (2.27)$$

which shows that for a scanned element contrast ratio greater than 10, D must be greater than 9 times the number of scanned rows. This leads to a severe requirement for D when a large number of rows is scanned, which may be difficult to achieve for some of the technologies, in particular the passive technologies such as LC, especially for the color displays. This has led to the development of the active-matrix addressing technique, discussed next.

2.4.3.2 Active-Matrix Addressing

As noted previously, the matrix-addressing technique termed "active" has been developed to overcome the difficulty in achieving acceptable contrast ratios when large arrays are scanned a row at a time. This is achieved by connecting a solid-state switch to each intersection point of the top and bottom conductor matrices. This technique was first developed by Brody [28] for use with EL displays, in the configuration shown in Figure 2.43, where the switching element is a thin-film transistor (TFT) as shown in Figure 2.44. The TFTs are then energized in the proper sequence at much more rapid rates than is possible with passive LC arrays, and can drive the elements to full saturation so that maximum contrast ratios can be achieved. However, this application has been superseded by its use for LC FPDs. It has been very successful in this embodiment, especially for color units, with amorphous silicon TFTs as the switching elements; the configuration shown schemati-

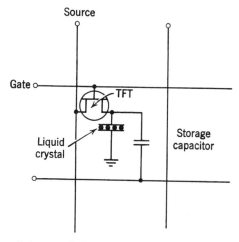

Figure 2.44 Design of elemental circuit for LC display. After Brody [28], by permission of SID.

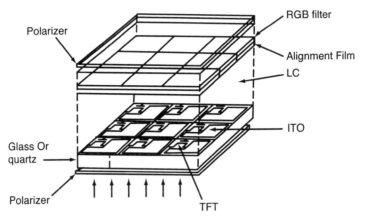

Figure 2.45   Active-matrix TFT LCD. After Schlam [29], by permission of SID.

cally in Figure 2.45 is one successful example of this technique. More detail on the complete assembly may be found in Chapter 3 in the section on monitors. At this point it is sufficient to note that active-matrix addressing has been applied with considerable success to a number of laptop and notebook color computers, and the technique remains the best approach to such product applications, at least until some of the other technologies, such as field emission, fully prove themselves.

2.4.3.3   Other Drive Techniques

A few other specialized drive techniques for matrix panels warrant some discussion at this point; further description is deferred to Chapter 3. These are the Self-Scan (registered trademark) techniques originally developed by Cola [14] for use with the dc plasma panels produced by Burroughs and now available from several other sources. Following the demise of Burroughs, the ISA technique was developed by Weber for use with ac plasma displays, with specialized grid and anode configurations for graphic dot-matrix VFDs. All of these techniques are used to simplify the drive circuitry for specific applications and have enhanced the success of these matrix panels in the marketplace. These drive circuitry technologies are described in later chapters with respect to the actual product for which they were designed. Other drive techniques may become available in the future, but these are the only ones used at present.

REFERENCES

1. Cornsweet, T. N., *Visual Perception*, Academic Press, New York, 1970, p. 40.
2. Kaufman, J. E., Ed., *IES Lighting Handbook*, 4th ed., Illuminating Engineering Society, New York, 1966.

3. (a) Farrell, J. E., et al., "Predicting Flicker Thresholds for Video Display Terminals," *SID Proc.* **28**/4, 1981, 451; (b) Sherr, S., *Fundamentals of Display System Design*, Wiley, New York, 1970.

4. Sherr, S., "Cathode Ray Devices," in J. Belzer, Ed., *Encyclopedia of Computer Science and Technology*, Marcel Dekker, New York, 1976.

5. Spangenberg, K. R., *Vacuum Tubes*, McGraw-Hill, New York, 1948.

6. Law, H. B., "A Three Gun Shadow Mask Color Kinescope," *Proc. IRE* **39**/10, 1951, 1186–1194.

7. Pender, H., and McIlwain, K., Eds., *Electrical Engineers' Handbook*, Vol. 2, *Communication-Electronics*, 4th ed., Wiley, New York, 1950.

8. Keller, P., "1976 CIE-UCS Chromaticity Diagram with Color Boundaries," *SID Proc.* **24**/4, 1983, 320.

9. Herold, E. W., "History and Development of the Color Picture Tube," *SID Proc.* **14**/4, 1974, 141–149.

10. Piper, W. W., and Williams, F. E., "Electroluminescence," in F. Seitz and D. Turnbull, Eds., *Solid State Physics*, Vol. 5, Academic Press, New York, 1957.

11. Goodman, L. A., "The Relative Merits of LEDs and LCDs," *SID Proc.* **16**/1, 1975.

12. Ivey, H. F., "Electroluminescence and Related Effects," in L. Martin, Ed., *Advances in Electronics and Electronic Physics*, Suppl. I, Academic Press, New York, 1963.

13. (a) Matsumoto, S., Ed., *Electron Display Devices*, Wiley, New York, 1990; (b) Karawada, H., and Ohshima, N., "DC EL Materials and Techniques for Flat Panel TV Display," *Proc. IEEE* **61**/7, 1973, 907–914.

14. Cola, R., et al., "Gas Discharge Panels with Internal Line Sequencing (Self Scan[(R)])," in B. Kazan, Ed., *Advances in Image Pickup and Display*, Academic Press, New York, 1977.

15. Nasser, E., *Fundamentals of Gaseous Ionization and Plasma Electronics*, Wiley, New York, 1977.

16. Jackson, R. N., and Johnson, D. E., "Gas Discharge Displays: A Critical Review," in L. Martin, Ed., *Advances in Electronics and Electronic Physics*, Vol. 35, Academic Press, New York, 1974, pp. 235–266.

17. Morimoto, K., and Pykosz, T. L., "Vacuum Fluorescent Displays," *SID Sem. Lect. Notes*, 1986, 1.2.20.

18. Schlam, E., "Overview of Flat Panel Displays," *SID Sem. Lect. Notes*, 1991, F-1/25.

19. (a) Goodman, L. A., "Passive Liquid Crystals, Electrophoretics, and Electrochromics," *SID Proc.* **17**/1, 1976, 30–38; (b) Goodman, L. A., "The Relative Merits of LEDs and LCDs," *SID Proc.* **16**/1, 1975, 8–19.

20. Pryce, D., "Liquid Crystal Displays," *EDN*, Oct. 12, 1989, 104.

21. Crossland, W. A., et al., "Prospects and Problems of Ferroelectric LCDs," *SID Proc.* **29**/3, 1988, 238.

22. Mannion, P., "Microtips Revitalize the CRT," *Electron. Proc.*, Dec. 1991, 17.

23. Kumar, N., et al., "Field Emission Displays Based on Diamond Thin Films," *SID 93 Dig.*, 1011.

24. Chan, C., "Diamonds in the Rough," *Broadcast Eng.* **1**/94, 18.

25. Deb, S. K., "Physics of Photochromic and Electrochromic Phenomena in Inorganic Solids," *Conference Record of 1974 Conference on Display Devices and Systems*, IEEE, SID, AGED, pp. 159–163.

26. Dalisa, A. L., "Electrophoretic Display Technology," *IEEE Trans. Electron Devices*, **ED-24**/7, 1977, 827–834.

27. Kmetz, A. R., "Matrix Addressing of Non-Emissive Displays," in A. R. Kmetz and F. K. von Willis, Eds., *Nonemissive Electrooptic Displays*, Plenum Press, New York, 1976.

28. Brody, T. P., "Large Scale Integration for Display Screens," *SID Proc.* **17**/1, 1976, 39–55.

29. Schlam, E., "Overview of Flat Panel Displays," *SID Sem. Lect. Notes*, 1991, F-1/30.

PART II

PRODUCT APPLICATIONS

3

OUTPUT DEVICES AND SYSTEMS

3.1 READOUTS

3.1.1 Introduction

The general designation of readouts refers to those products that provide displays in the form of numerics and alphanumerics, now almost exclusively using one of the flat-panel technologies covered in Chapter 2. The most popular technologies employed for this display applications are listed in Table 3.1, and it should be noted that EL has no sources except for backlighting, and ac plasma has only one vendor that offers a line of such products using ac plasma technology, but with only minimal acceptance.

Readouts have a broad and varied range of applications, from simple watch and calculator displays to more complex instrument displays that may have A/N assemblies as well as bar graphs. In any event, this is a significant product area, and warrants considerable attention and description of the various technologies and formats in which readouts may be found. These are covered in the following sections.

3.1.2 Light-Emitting Diodes (LEDs)

LEDs as display elements may be found in either monolithic or discrete forms, depending on the size of the character. The first form consists of a multiplicity of elements on a single substrate, as is shown in Figure 3.1, which demonstrates the construction of a LED monolithic chip consisting of a single seven-segment numeric. This structure can be extended to contain a number of digits or larger numbers of LEDs in matrix form, leading to many different assemblies in a variety of forms as described later in this chapter. This structure is also used for bar graphs in which up to 200 individual bars are placed in a stacked array on a single chip.

TABLE 3.1   Readout Technologies

Technology	Types	Comments
LED	Numeric, A/N	Generally limited to 4–8 characters
EL	Backlight	Not used for readouts
Plasma	Numeric, A/N, bar graph	Dc more common than ac
VF	Numeric, A/N, bar graph	Used for indoor equipment
LC	Numeric, A/N, bar graph	Most popular for portable applications
EM	Numeric, A/N	Most popular for outdoor applications
IN	Numeric, A/N	Limited outdoor use

The other form in which LED displays may be found—namely, the one that uses discrete LEDs—appears in a large variety of formats, ranging from single numerics to large boards containing thousands of individual units. The individual LEDs are arranged in assemblies of matrices as shown in Figure 3.2*a* or in large matrix arrays as shown in Figure 3.2*b*, which can be used as message boards with some graphics capabilities. These latter applications have been somewhat limited because of the relatively high current requirements when a fairly large number of LEDs are used, but this limitation has been considerably reduced because of the more efficient LEDs that have become available. Thus, it may be anticipated that LED readouts may be used more extensively for outdoor applications. A representative set of parameters for this type of discrete LED display is given in Table 3.2.

3.1.3   Electroluminescence and Plasma

The next technologies of interest are other light-emitting types, namely, electroluminescence (EL) and plasma. In the first example, the light-emitting element could consist of groups of individual light emitters, somewhat similar in size to the monolithic LEDs, and arranged in a variety of arrays, or as larger panels that can be used for backlighting of displays using a non-light-emitting technology such as liquid crystals. EL is an old technology in its earlier field emission form, and some readouts were sold in this form but EL has been most successful for information displays primarily in its thin-film (TFEL) form originally developed by Sharp, and described in Section 2.3.3.3, with the basic structure used for this technology shown in Figure 2.22. However, whereas this technology has been used to create A/N and graphics panels of various sizes, it has not been successfully designed to produce the simpler readouts. Therefore, further discussion of this technology is deferred to later sections in this chapter.

As to plasma displays, the same may be said for ac types in that only one vendor has offered readouts using ac plasma technology, although there are a number of sources for the dc versions. These latter may be found in two basic versions, the first termed *planar*, which is appropriate for a limited number of characters; and *Self-Scan*, which is more appropriate for larger arrays.

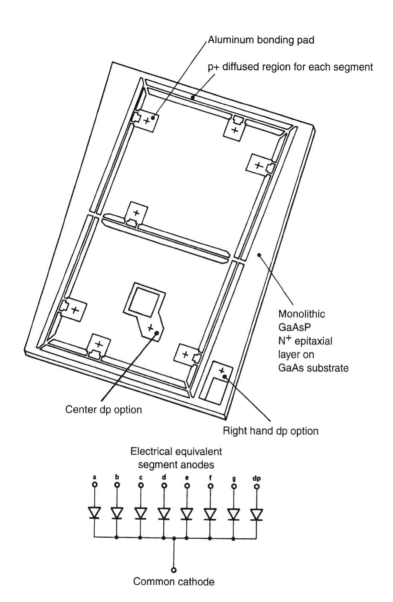

Aluminum bonding pad

p+ diffused region for each segment

Monolithic
GaAsP
N$^+$ epitaxial
layer on
GaAs substrate

Center dp option

Right hand dp option

Electrical equivalent
segment anodes

a b c d e f g dp

Common cathode

Figure 3.1 Construction of LED monolithic chip. After Gage et al. [1]. Permission to reproduce granted by Hewlett-Packard Co. Copyright 1981.

Figure 3.2 (*a*) Photograph of LED matrices. Courtesy of Total Control. (*b*) Photographs of large LED matrices. Artwork courtesy of Daktronics, Inc.

TABLE 3.2 Large Discrete LED Digit Display Parameters

Parameter	Value
LED size (in.)	0.2 diameter
Character form	2 in. = 5 × 7 dot matrix
	4 in. = 10 × 14 dot matrix
	6 in. = 15 × 21 dot matrix
Display field (in.)	9.2 × 30.8–98 × 242
Characters per line	6–80
Number of lines	1–40
Matrix ($H \times W$)	24 × 96–192 × 480

Source: Gage et al. [1].

Self-Scan is also designated as raised-cathode and screened-image, depending on the exact method of construction. The basic structure of a dc panel in its simplest form, as shown in Figure 2.23a, consists of three insulated plates sealed at the circumference, with the outer two containing conducting electrodes in the vertical and horizontal directions on the inside of the glass, and the center plate made with a matrix of etched holes. A neon-type gas is inserted between the two outer plates, and if the gas is made to go into the discharge mode by the application of a dc field across the plates, then the discharged gas will emit light of an orange color, similar to that seen when the well-known Nixie tube is activated. The discharge and subsequent light emission occurs only between the electrodes that have the field between them, and is limited to the intersection of the electrodes. Thus, the light emission can be made to take on the form of the electrodes in either dot-matrix or line form, and thus create the desired characters by selecting the combination of dots or lines required to delineate the characters. Figure 3.3a demonstrates the steps in the construction of a screened image type of multicharacter assembly, and Figure 3.3b is a photograph of a representative display module.

Proceeding next to the raised-cathode type, the structure is shown schematically in Figure 3.4a, and has the advantage of achieving high levels of luminance, although it is usually limited to rather simple numerics to achieve these results. However, it has found a fairly large market as the preferred display for various outdoor applications such as gasoline pumps, toll booths, and outdoor scales. Characteristic operating parameters for either of these two planar types are given in Table 3.3. It is of some interest to examine a characteristic driver circuit as shown in Figure 3.4b as it demonstrates the potential complexity of such circuitry when more than a few characters are involved with the necessity for multiplexing when any significant number of characters are present. This leads in turn to the Self-Scan version covered next that has as its main advantage the simplification of the driver circuitry at the cost of increased complexity in the physical construction.

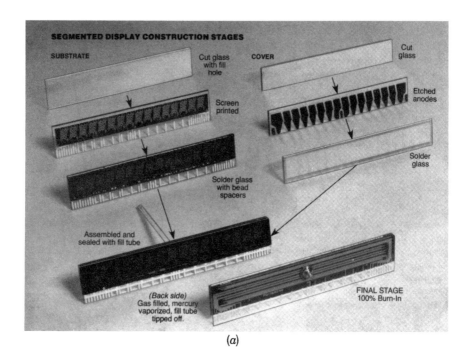

(a)

(b)

Figure 3.3 (a) Direct-current plasma panel construction stages. Courtesy of Dale Electronics, Inc. (b) Photograph of plasma display module. Courtesy of Cherry Electrical Products.

The Self-Scan type of dc plasma readout was originally invented at the former Burroughs Corporation, and subsequently taken over by several other companies after the demise of Burroughs. This approach had been used for both limited A/N units and full matrix panels, but at present only the bar-graph types remain, although the same technical considerations hold for all the types. A simplified diagram of the electrodes for a representative panel is shown in Figure 3.5a. The unique feature of this type of structure is that addressing and scanning are achieved by means of priming and transfer of the priming signal sequentionally from the first position to the following ones. This is accomplished by establishing a glow in the gas at the first position on the bottom side of the panel shown in Figure 3.5a. This glow then appears at the top side that contains the display anode and cathode, and is transferred along by the priming signal moving through small holes from the priming section to each of the other sections, sequentially. This transfer of ignition from the first section to the last section in a row is done without the need for electronic scanning or multiplexing signals, so that the electronics and connections are considerably reduced by using this technique. It is most effective for bar graphs with up to 2 bars of 201 elements each, although the electronics is somewhat complex as shown in Figure 3.5b, requiring 3-phase

Display construction

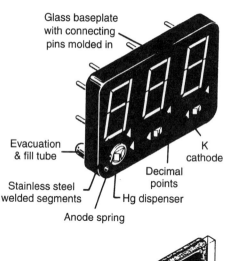

Glass baseplate with connecting pins molded in

Evacuation & fill tube

K cathode

Decimal points

Stainless steel welded segments

Hg dispenser

Anode spring

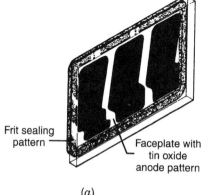

Frit sealing pattern

Faceplate with tin oxide anode pattern

(a)

Figure 3.4 (a) Plasma panel raised-cathode structure.

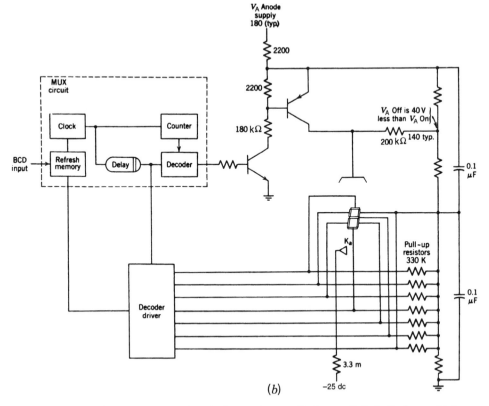

Figure 3.4 (*b*) Block schematic of planar dc gas discharge digital display system. Courtesy of Babcock Display Products Group.

circuitry as well as the anode drivers. However, the resulting images can be quite effective as shown in the photographs of Figure 3.5*c*. A characteristic set of parameter values for the bar-graph types is given in Table 3.4.

The bar-graph array has seen some success as an instrument readout, although initial attempts to use it for automotive displays have been relatively unsuccessful due to electrical noise generated by the display and the rather low luminance achieved.

TABLE 3.3   Planar DC Plasma Display Parameters

Parameter	Value
Number of characters	4–32
Character height (in.)	0.25–2
Luminance (fL)[a]	60–150

[a] Foot-lamberts.

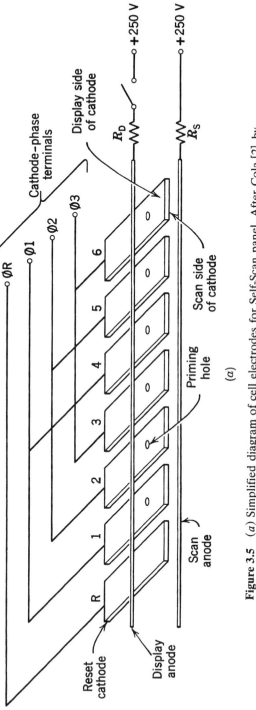

Figure 3.5 (*a*) Simplified diagram of cell electrodes for Self-Scan panel. After Cola [2], by permission of Academic Press.

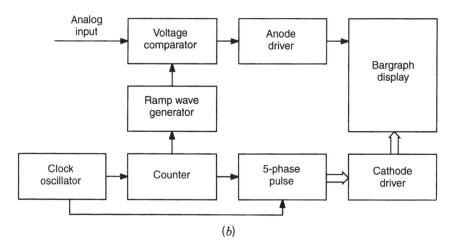

(b)

(c)

Figure 3.5 (*Continued*) (*b*) Block schematic of bar graph system. Courtesy of Babcock Display Products Group. (*c*) Photograph of bar graph display. Courtesy of Babcock Display Products Group.

TABLE 3.4   Bar-Graph Parameters

Parameter	Value
Number of elements	101, 201
Resolution (%)	0.5, 1.0
Segment length (in.)	0.1-0.25
Segment height (in.)	0.01-0.025
Luminance (fL)	30-70
Refresh rate (Hz)	70

3.1.4   Vacuum Fluorescence

Vacuum fluorescent displays have been very successful as readouts in the numeric, A/N, and bar-graph formats. They are available in both segmented and dot-matrix versions, both as displays without drive electronics and as complete operating modules with self-contained character generating, multiplexing, and control capabilities. The block diagram for one such module, shown in Figure 3.6a, represents the characteristic complexity of the associated electronics. This circuitry could be somewhat costly if not reduced to a single chip, so it is customary to sell the unit as a complete module to reduce the cost. In addition, different circuitry is required for the segmented and dot-matrix versions so that it is generally impractical for the user to buy the display without the circuitry unless the quantities are large enough to warrant developing the required microcircuit chip. Therefore, the majority of the VF readout units are self-contained, with all of the required circuitry for driving and multiplexing, and may be used directly as operating displays without the addition of any other circuitry. This makes VFDs quite attractive, especially for small displays, and they are found in a number of products that use such displays, of which audio, video, and radio receivers are characteristic examples. These units employ the structures shown in Figure 2.25, and, as previously noted, are basically flat CRTs. Thus, $B+$ and filament voltages are required, but the relatively low $B+$ voltage of 25 V dc and low filament power, make these units quite satisfactory for these representative applications. The operating parameters for several different types of these readout units are given in Table 3.5 in terms of their specific configurations.

3.1.5   Liquid Crystal

Liquid crystal displays are probably the most successful and ubiquitous examples of readout displays, in particular for portable and handheld units. They first made their mark using the twisted nematic technology, but this technology has been superseded by the supertwisted and double-supertwisted versions. For a time, the earlier types threatened to take over the watch display application, but with the return of the analog readout versions, the LC types have been more or less relegated to a subsidiary role. However, for

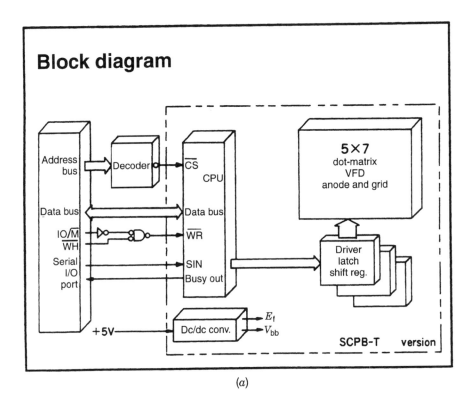

(a)

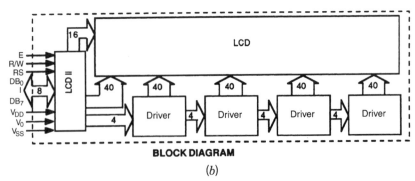

BLOCK DIAGRAM

(b)

Figure 3.6 (a) Block diagram of vacuum fluorescent module. Courtesy of Noritake Co. Inc. (b) Block diagram of 5 × 7 dot matrix LCD module. Courtesy of Standish.

TABLE 3.5 Vacuum-Fluorescent A / N Display Parameters

Parameter	Type	Value
Number of lines	14/16 segments	1–2
Number of characters	—	2–20
Number of lines	5 × 7 dots	1–2
Number of characters	—	2–40
Number of bars	Bar graph	1–2
Number of segments	—	51–201
Character size ($H \times W$, mm)	14/16 segments	5 × 3–13 × 7.8
Character size ($H \times W$, mm)	5 × 7 dots	5 × 3–15.1 × 18.2
Bar size ($H \times W$, mm)	—	5 × 50.4–8 × 100
Luminance (fL)	Segments, dots	60–200

TABLE 3.6 Applications for LCD Readouts

Application	Type	Application	Type	Application	Type
Multimeter	Numeric	Printer	A/N	Organizer	A/N
Calculator	Numeric	Cash register	A/N	Scale	Numeric
Telephone	Numeric	Audio/video	Numeric	Pager	Numeric
Facsimile	A/N	Camcorder	A/N	Gasoline pump	Numeric

battery-operated devices such as multimeters and calculators, and a host of other specialized units such as organizers, telephones, and facsimile machines, LCD numeric and A/N readouts predominate. Indeed, for many of the applications requiring readouts, LCDs have become the preferred device, especially when battery operation and low power requirements are important considerations. A more complete list of the applications for LCD readouts is given in Table 3.6, and demonstrates the wide range of uses for this type of display product.

It is evident that the number of these applications for LCD readouts is quite large, and it should be noted that many of these application examples may also use some of the other technologies, when battery operation is not essential. Some parameter values for LCD readouts of A/N dot-matrix types are given in Table 3.7, and a block diagram for a 40-character module is shown in Figure 3.6b.

TABLE 3.7 Dot-Matrix LCD Parameters

Parameter	Value
Characters × lines	16 × 1–40 × 2
Character height (mm)	3.3–12.7
Viewing area $H \times W$ (mm)	15 × 35–68 × 245

3.1.6   Other Technologies

Some of the other technologies have found rather limited acceptance for readout applications. Among these, the most successful have been incandescent (IN) and electromagnetic (EM), with the latter found primarily in outdoor locations such as toll booths, or indoors where large A/N signs and message displays are found. However, these are rather limited applications and are mentioned here primarily for completeness. Electrochromics and electrophoretics have also made sporadic efforts to find a market but without much success.

3.1.7   Technology Comparisons

To conclude this section on readout applications, it seems advisable to provide some means for comparing the advantages and disadvantages of the main technologies for readout applications. The goal of this effort is to aid the user in choosing the best or at least the most appropriate technology and device type for any particular application. In some cases this might be very difficult, as each technology has certain advantages and disadvantages that must be related to the specific requirement of the application. Thus, at best the choice remains tentative, but this approach at least is better than none at all. This comparison is done in Table 3.8, which should offer some means for

TABLE 3.8   Advantages and Disadvantages

LED	Plasma	Fluorescent	LCD
Advantages			
Wide temperature range	Appearance	Low voltage	Very low voltage
Very low voltage	Long Life	Low power	Very low power
Color	Color	Color	Color
Long life	Fast response	Moderate response	Sun readable
Multiplex capable	Multiplex	Multiplex	Passive matrix
Matrix-addressable	Matrix	Matrix	Active matrix
Very fast response	Fast response	Fast response	Large size
Disadvantages			
Limited number of characters	High voltage	Vacuum bottle	Nonemissive
High power	Moderate luminance	Filament heater	Limited angle of view
Limited size	Moderate contrast ratio	Fragile	Limited temperature range
Multiple elements	Special drivers	—	Special matrix required

making selection among these technologies in terms of their major operational characteristics.

3.2 PANELS

3.2.1 Introduction

Panel displays are available in a variety of formats, ranging from relatively simple three- or four-line units with up to 40 characters a line, to highly complex graphics displays with high-resolution capabilities. The technologies are the same as those used for readouts, plus EL and ac plasma as full participants, with the addition of flat CRTs as part of this group. These are listed in Table 3.9. The applications for these FPDs are extremely broad, including indoor and outdoor message and advertising displays, a variety of graphic displays, and most significantly, display panels for the many different types of portable, battery-operated computers, measuring instruments, oscilloscopes, TVs, and games that have proliferated with the introduction of improved versions of displays using the five major technologies for FPDs used for these applications.

3.2.2 Light-Emitting Diodes (LEDs)

The LED types used for the larger panels are restricted to the discrete units, which may be assembled into modules, or installed as individual units on panels of various sized, and addressed in matrix fashion. The major application is for large board A/N displays, although occasionally simple graphic displays such as that shown in Figure 3.7 can be found. The discrete diodes are arranged in matrix form, and are addressed in the usual matrix manner. Another form in which these FPDs can be constructed is by using basic modules with 16×16 or other matrices of either discrete or monolithic variety, for which one representative drive circuit is shown in Figure 3.8. Assemblies of these modules can be arranged so that the equivalent of a 640×480 or better display can be obtained, which makes it possible to

TABLE 3.9 FPD Technologies

Technology	Types	Comments
LED	Message, graphic	Discrete LEDs, matrix-addressed
EL	Graphic, dc, ac	Matrix-addressed panels
Plasma	A/N, graphic, dc, ac	Matrix-addressed ac, scan dc
VF	A/N, graphic	Matrix-addressed
LC	A/N, graphic	Matrix-addressed, active, passive
FE	Message, graphic	Matrix-addressed, cold cathode
EM, IN	Message A/N, graphic	Matrix-addressed, discrete bulbs

display graphics images of some complexity on this type of LED display unit. However, color remains a problem, although the development of viable blue-emitting units has made it feasible to obtain full-color graphic displays. Parameters of such modules are given in Table 3.10. However, it should be noted that since a full-color display requires at least three separate diodes for each matrix element of the total display, this will increase the complexity, cost, and power requirement by considerable amounts. For this reason, LED FPDs have tended to be restricted to single-color units and to large boards with limited numbers of A/N characters.

3.2.3 Electroluminescence (EL)

3.2.3.1 Alternating Current (AC)

With EL technology we come to the first truly graphic display panels of the group of major FPD technologies, in spite of the capability of limited graphics on some of the LED units previously described. Table 2.10 lists four different types of EL technologies (ac powder, film; dc powder, film) that have seen some use as display media. However, of these, only dc powder and ac film have shown any significant market success, and therefore the discussion at this point is limited to those two, beginning with TFEL as the ac version. As noted previously, this technology was first successfully developed by Sharp, and for a while it was the major supplier of the panels made using this technology. However, it was soon followed by Planar, which, with the acquisition of the only other major source, that is, Lohja in Finland, or Finlux in the United States, became the leader in this field. Planar has continued to improve and further develop TFEL panels, having introduced a new manufacturing technique it terms *integral contrast enhancement* (ICE™), which improves the contrast ratio and character sharpness. This is accomplished by modifying the row electrode as is shown in Figure 3.9, where the new technique is compared with the earlier approach. The improvement is achieved by depositing a gradient-index film, that absorbs the ambient light rather than reflecting it, between the dielectric layer and the row electrode. Other advances are the full-color units in two configurations, one using a white phosphor with color filters as shown in Figure 3.10a and the other based on the development of a blue phosphor, in combination with red and green phosphors, as illustrated in Figure 3.10b. Full-color units employing the second approach are available in a range of sizes and resolution from Planar. Competitive developments from Sharp are the unit with a 17-in. diagonal and the high-resolution (1280 × 1024) 13-in. diagonal unit, as well as the high luminances made available by using the split-screen drive shown in Figure 3.11, where the dwell time is doubled, thus achieving higher luminance. This technique is also used by Planar in an 18-in. diagonal unit with 1024 × 864 resolution. Finally, a high-resolution active-matrix EL display has been announced by Planar in conjunction with several other organizations [5]; the basic construction is shown in Figure 3.12a, where it is

7x80 points

16x80 points

24x80 points

32x80 points

40x96 points

40x96 points

40x96 points

Figure 3.7   LED graphic displays. Artwork courtesy of Daktronics, Inc.

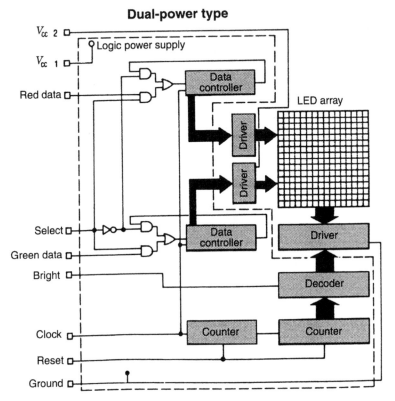

Figure 3.8 Block diagram for LED 16 × 16 dot-matrix display module. Courtesy of Toshiba.

TABLE 3.10 LED Module Parameters

Parameter	Value
Luminance/dot (nits)	50–190
Colors	R, G, B
Module matrix	4 × 4–16 × 16
Current/LEDs (A)	1.3–2.4
Module size (mm²)	63.7–95.7

compared with a standard structure. The active-matrix EL (AMEL) has better than 1000-line in. resolution with luminance above 500 fL and contrast ratio greater than 100. The pixel circuit design for the active matrix is shown in Figure 3.12*b*. All of these advances place TFEL in the forefront of FPDs. In addition, Planar has a group of monitors available, but discussion of these units is deferred to Section 3.3.

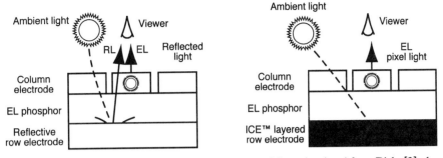

Figure 3.9 Electrode modification for EL row drive circuit. After Birk [3], by permission of SID.

3.2.3.2 Direct Current

Direct-current is of more recent vintage than TFEL, but not very far behind. It originated in England at Phosphor Products, but has been taken over by Cherry in the United States; Cherry was responsible for its more recent development and marketing. However, Cherry has ceased to offer dc EL products, so they are probably defunct for now, although they may reemerge, as has been the case for other technologies. The similarity between the structures for the ac and dc versions is illustrated in Figure 3.13. The major difference is that in the dc version the phosphor layer must be connected to the drive directly on both sides, whereas the ac type requires dielectric layers between the drive and the phosphor layer. Apart from this difference, the two approaches are essentially identical. Thus, the success of ac EL may make the dc technology viable again, as the number of suppliers of ac units remains rather limited at present and the potential for an increase appears rather low. In the event that the demand for ac EL increases, the dc versions could provide adequate performance as well for some applications, and a set of parameters is given in Table 3.11 for historical if not practical purposes.

3.2.4 Plasma

3.2.4.1 Direct Current (DC)

Dc plasma is the earliest form of FPD using plasma technology, going back to the earliest Nixie readouts, which, while not FPDs, still instituted the use of gas discharge for display devices. From this beginning, Burroughs proceeded to displays with larger numbers of characters, culminating in the Self-Scan version that established a basis for plasma FPDs that was taken over by a number of other manufacturers after Burroughs left the field, and had some impact on the later dc plasma designs that followed. However, the latest versions of Self-Scan remain viable only in the form used for the bar-graph types, described in Section 3.1.3, with no capability for characters or any extensive matrix display capability, and have been supplanted by other techniques for large numbers of characters in matrix-addressed formats. An

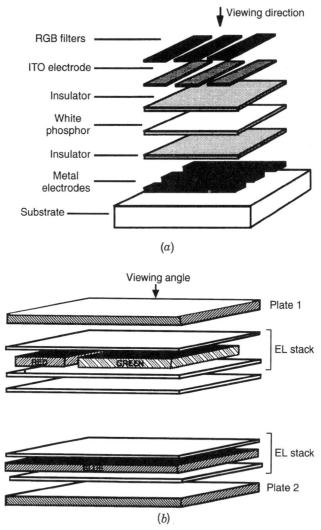

Figure 3.10 (*a*) Color filter structure for TFEL display. (*b*) Dual-substrate structure for color EL display. After King [4], by permission of SID.

example is a unit that displays 8 lines of 32 characters in a 5 × 7 dot matrix, with the parameters are given in Table 3.12.

In addition to these, several other types of dc plasma displays have been devised using direct-driven matrix structures that have achieved somewhat greater matrix capabilities, including at least one with color and HDTV capability, for which representative parameters are listed in Table 3.13. This is a 33-in. diagonal unit developed by NHK. However, it is too soon to

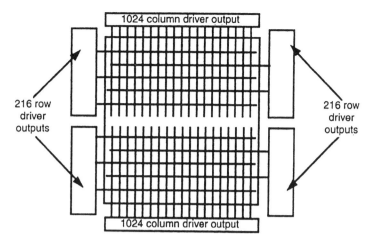

Figure 3.11   Split column electrode structure for EL matrix display. After King [4], by permission of SID.

predict its success unit more units are in use, and the practicability of LC projection units has been fully established.

The electrode structure of the dc plasma display is shown in Figure 3.14, and consists of a front and rear glass with each color pixel having four cells made up of one red, one blue, and two green cells. A simpler design, more practical for the majority of FPD applications, is matrix-addressed and driven with the block diagram shown in Figure 3.15. This type of FPD appears to be practical for desktop computers, but requires too much power to be applicable to laptop or other battery-operated units. However, good luminances and contrast ratios can be achieved, as is shown by the parameters for this design, listed in Table 3.14 for the refreshed dot-matrix panel.

This is quite a respectable unit for applications that can live with monochromatic displays, but since it finds considerable competition from full-color CRT displays, efforts have been made to produce color versions of these dc plasma panels with some success, although improvements in luminance are still necessary to make it fully viable.

3.2.4.2   *Alternating Current (AC)*

Next we come to the plasma panels that use ac drive and are essentially restricted to applications that use panels with various size matrices of display elements, up to as large as 2048 × 2048 pixels in a panel with a viewing area of 31.4 × 23.8 in. The range of these parameters and others for ac plasma panels is given in Table 3.15.

A block diagram of a circuit for a panel with a resolution of 1024 × 768, shown in Figure 3.16, has the potential for use as the display for HDTV, at least in the monochromatic (orange) color, but the latest developments in ac plasma FPDs have gone beyond this point by adding full color. These include

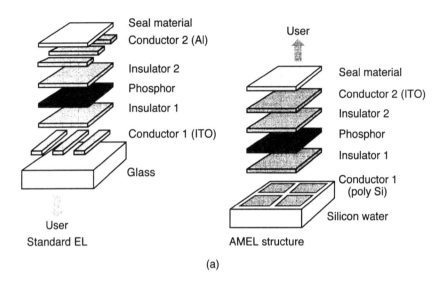

(a)

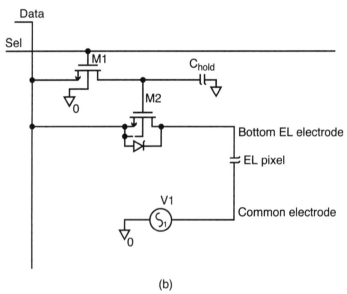

(b)

Figure 3.12 (*a*) Comparison of standard and active matrix EL structures. (*b*) Pixel circuit design for active matrix EL display. After Khormael et al. [5], by permission of SID.

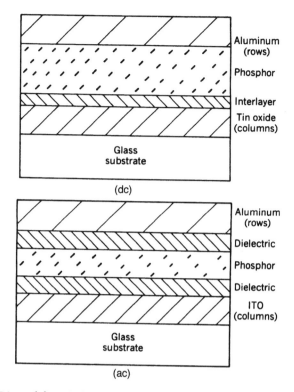

Figure 3.13 Direct-(*a*) and alternating-current. (*b*) ac EL panel construction. After Vecht [6], by permission of SID.

TABLE 3.11   DCEL Panel Parameters

Parameter	Value
Luminance (fL)	25–30
Active area (in.)	3.68 × 2.94–8.64 × 6.48
Matrix ($W \times H$)	320 × 250–640 × 480
Color	Amber
Contrast ratio	15–20

TABLE 3.12   Optical Parameters

Parameter	Value
Character height (in.)	0.34
Character spacing (in.)	0.22
Character format	5 × 7
Viewing angle	130°
Luminance (fL)	20–50

TABLE 3.13   HDTV DC Color Plasma Panel

Parameter	Value
Size (mm)	520 × 665
Number of cells ($H \times W$)	800 × 1024
Luminance (nits)	85
Contrast ratio (no ambient light)	820

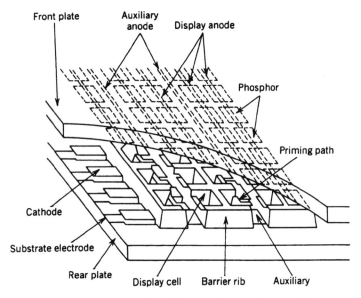

Figure 3.14   Electrode structure of NHK's dc plasma display. After Uchike [7], by permission of SID.

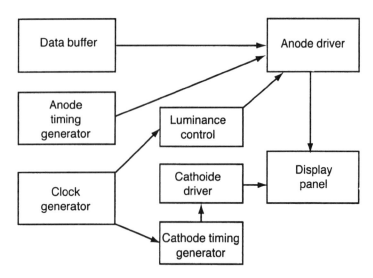

Figure 3.15   Block diagram of dc plasma display unit.

TABLE 3.14   Refreshed DC Dot-Matrix Plasma Panel

Parameter	Value
Viewing area (mm)	89×195–160×256
Matrix ($W \times H$)	88×192–480×640
Luminance (nits)	50–250
Contrast ratio	10–150

TABLE 3.15   AC Plasma Panel

Parameter	Value
Viewing area (in.)	4×4–32×24
Matrix ($H \times W$)	240×80–2048×2048
Resolution (lines/in.)	50–85
Luminance (fL)	6–85
Contrast ratio	10–20
Color	Neon orange

those developed by Fujitsu (see structural details in Fig. 3.17a), by Thomson CSF (structural details in Fig. 3.17b), and others from Photonics and Plasmaco. Color is achieved, as shown in these illustrations, by adding red, green, and blue phosphor elements to the panels and activating the phosphors by either UV or ionic bombardment from the plasma, depending on which structure is used. Relevant parameters for these units are given in Table 3.16. The importance of color is evident, and the availability of these units has brought ac plasma back into the market for FPDs that may be used for HDTV.

3.2.5   Vacuum Fluorescence (VF)

VFDs have achieved their greatest success in applications requiring readouts of the numeric, A/N, and bar-graph varieties, but they have been somewhat rejected as A/N and graphics panels. The main deterrent to full acceptance appears to be the limitation on maximum size, which is in the order of 7.1×6.3 in., although a resolution of 640×400 is available at this size. In addition, different colors can be provided by using filters, but only one at a time. However, this limitations is somewhat compensated for by the very high luminances that can be attained of over 400 fL in certain configurations. Finally, there are several unique construction and drive techniques available for graphics displays, in addition to the simplest single-matrix approach. These are the dual-wire-grid scanning and the triple-matrix anode techniques that reduce production costs and provide improved visual appearances.

The dual-wire technique for a single-matrix system is illustrated in Figure 3.18a, which shows the electrode structure and timing chart of the grid and

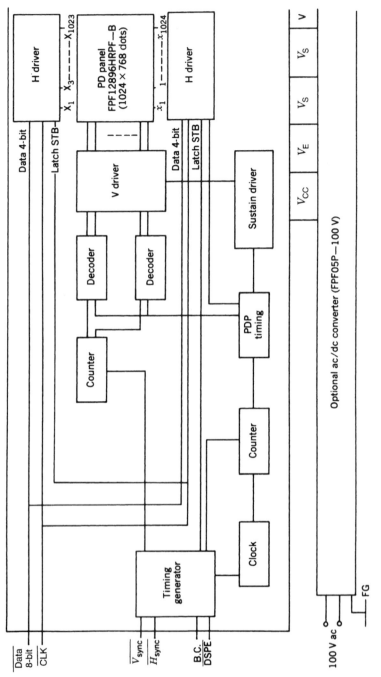

Figure 3.16 Block diagram for 1024 × 768 matrix ac plasma display. Courtesy of Fujitsu.

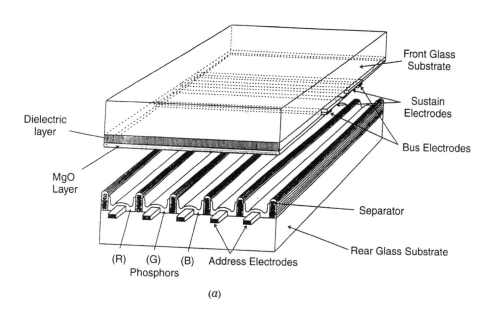

(a)

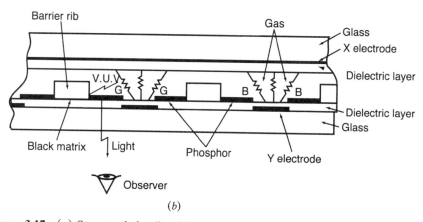

(b)

Figure 3.17 (a) Structural details of Fujitsu ac plasma panel. Courtesy of Fujitsu. (b) Structure of color ac panel developed by Thomson. After Weber [8a], by permission of SID.

TABLE 3.16   Optical Parameters for AC PDPs

Company	Diagonal (in.)	Matrix ($H \times W$)	Colors (No.)
Fujitsu	21	640 × 480	18 bits
Plasmaco	21	640 × 470	16 million
Photonics	30	1024 × 768	18 bits
Photonics	19	640 × 480	256,000
Thomson	13	640 × 480	256,000

anode for dual-wire scanning. In this structure the anodes are formed as stripes and the grid wire is placed at right angles, which is fairly standard for matrix displays. However, it differs in that a set of two adjacent grid wires is shifted sequentially and turned on when synchronized "on" and "off" signals are applied to the anode. Then as shown in the timing chart, the anode that is between the grid pair is activated and light is emitted at that point. This technique has the advantage that lower tolerances are needed for the anode construction, leading to reduced manufacturing cost, and vertical, horizontal, and oblique lines all appear to be continuous, leading to improved appearance. The triple-anode construction is considerably more complex, with anode connections made in three matrices and the grids arranged to cover two anodes, as shown in Figure 3.18b, so that although the anode connections are triple the number of dots in a vertical line, the number of grid

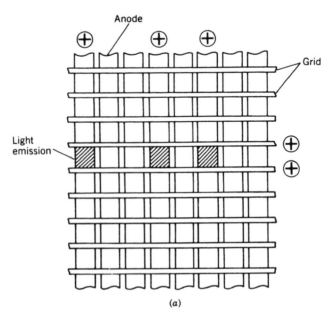

(a)

Figure 3.18   (a) Electrode structure of dual-wire scanning VFD.

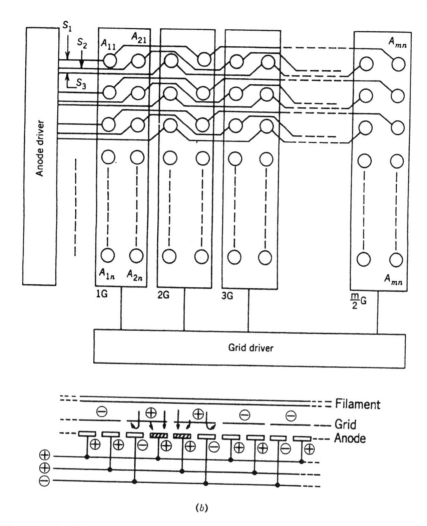

Figure 3.18 (*b*) Operational principle of triple-anode VFD. Courtesy of Futaba.

connections is half that in a horizontal line. This arrangement allows higher luminances to be achieved by repeating the duty cycle, as two dots are covered by one grid, and lower voltage operation is possible, thus compensating to some extent for the increase in the number of connections. Other features of VFDs are found in one version that accepts RGB signals, which, with the use of an LCD color shutter, allows three colors to be displayed. One approach has been the addition of active-matrix drive with each pixel a phosphor layer on the drain electrode of a metal oxide semiconductor field-effect transistor (MOSFET). Although not too successful as yet, this and

TABLE 3.17   VFD Parameters

Parameter	Value
Display area (mm)	$48 \times 25 – 179 \times 112$
Matrix ($H \times W$)	$64 \times 32 – 640 \times 400$
Luminance (nits)	50–690
Dot size (mm)	0.22–0.65

other attempts may lead to panels with more capabilities and a wider acceptance for applications beyond those for which the readout types are adequate. Until then, some representative parameter values are shown in Table 3.17. Within the limits of these parameter values, VFDs make effective and attractive FPDs, and can be considered for a variety of applications.

3.2.6   Liquid Crystal Displays (LCDs)

3.2.6.1   Supertwist

The supertwist and double-supertwist types of LCDs have essentially superseded all earlier versions, and are the only ones used in the panel displays considered here. The basic structure is shown in Figure 2.30, and the mode of operation is shown in a more detailed form in Figure 3.19, where the compensator cell is replaced by a polymer film that improves the contrast ratio. Parameters for this structure are given in Table 3.18, and it may be assembled in large arrays, for which a block diagram of a drive unit is shown in Figure 3.20.

Larger assemblies with resolutions as high as 1280×1024 pixels have also been constructed and are available for applications that require such resolution, such as HDTV and some computer types. Color may be achieved by the use of three color polarizers as illustrated in Figure 3.21, and for which a set of parameter values is given in Table 3.19. These are only representative, and more colors and hogher or lower resolutions are attainable, although the structures become rather complex, and special matrix drive techniques must be used. These parameters can be improved by going to the more complex matrix addressing and drive techniques such as those termed "active" and "passive" matrix covered in the next section.

3.2.6.2   Active- and Passive-Matrix Addressing

Matrix addressing is covered in some detail in Section 2.4.3, but the special features of the so-called active-and-passive matrix techniques are given only limited attention there. However, as these techniques are of prime importance in making LCD panels, especially the high-resolution color types, feasible and practical, it is advisable to describe them in more detail in the context of the specific panel types. Matrix addressing and driving is the characteristic way in which FPDs are activated and in its simplest form

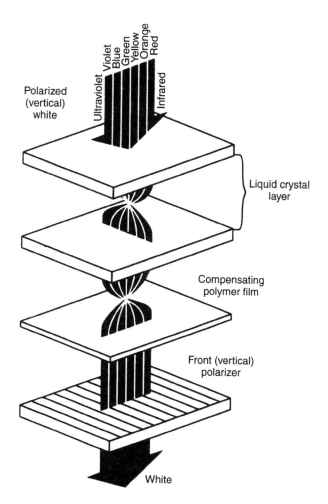

Figure 3.19 Structure of compensated supertwist LCD using polymer film. After Lieberman [8b], by permission of *Computer Design* (April 1989, p. 37).

TABLE 3.18   STN LCD Parameters

Parameters	Value
Viewing area (mm)	46 × 18–220 × 166
Number of pixels	32 × 80–480 × 620
Dot size (mm)	0.24–0.5
Response time (ms)	100–300
Contrast ratio	5–15

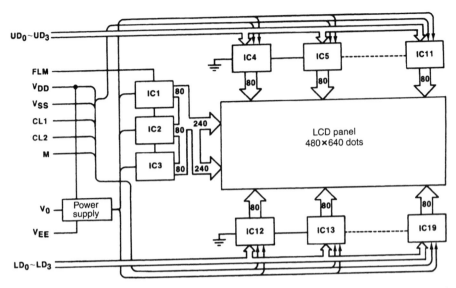

Figure 3.20   Block diagram of 480×640 supertwist LCD panel. Courtesy of Densitron.

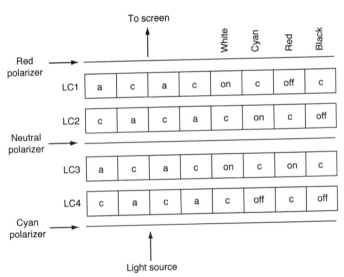

Figure 3.21   Multiple high-resolution display schematic. After Conner and Gulick [9], by permission of SID.

TABLE 3.19   **Color STN Display Parameters**

Parameter	Value
Display area	182×140–250×187
Number of pixels	1920×400–1920×480
Colors (no.)	8–256
Contrast ratio	10–20
Response time (ms)	100–200

consists of the X–Y square array shown in Figure 2.41, and discussed to some extent in Section 2.4.3. This approach to the selection and energizing of elements in matrix panels is used extensively for the various FPD technologies discussed previously, but in the case of the LCD types, some variants on the basic matrix approach have evolved to cope with some of the limitations of the LC material, in particular the excessive response times of the earlier materials. This response time remains a significant problem, although it is being somewhat ameliorated by new materials, and the techniques of active- and passive-matrix addressing are being used to overcome the deficiencies resulting from this limitation. These techniques are discussed next, beginning with the first to achieve considerable success, that is, active matrix addressing.

As noted previously (Section 2.4.3.2), active-matrix addressing was originally developed for EL displays, but had its greatest success initially with LC FPDs. To this statement, we can add that supertwist and double-supertwist versions have taken over the market for active-matrix although some EL vendors are attempting a comeback for active-matrix-driven EL panels. However, the major application for active-matrix addressing remains the drive circuitry for active-matrix LCD (AMLCD) panels. These panels have achieved impressive results, especially in the development and manufacture of FPDs with a 256-color capability as a tentative standard for computer displays, with 4096 colors readily attainable, and as many as 16.7 million colors not far behind. Thus, full-color panels for HDTV are not far away, and panels with 1280×1024 resolution are being offered, with the maximum size in the neighborhood of a 16-in. diagonal, although the response time is still not completely satisfactory for a noninterlaced, high-resolution display. However, improved drive techniques promise to bring the pixel response time down to 20 ms or less, which should be acceptable for a TV display. However, the size limitation appears to be more intractable, with projection panels as the most likely solution. This is discussed further in Section 3.5.

Passive-matrix addressing (PMA) has come along quite rapidly as an alternative to active-matrix addressing (AMA), for which the costs have tended to remain high. There may be a little semantic confusion where passive-matrix addressing is concerned, because it was originally referred to as "active addressing," which is very similar to "active-matrix addressing," but the technique is quite different from the latter approach. The major difference is that whereas the former required the addition of some type of active driver assembly to improve the response time of the system, the latter achieves its somewhat similar results by a judicious use of drive waveforms. It differs from standard matrix addressing in that instead of using rows driven sequentially within a frame time with the corresponding column signals driven in synchrony, a specially designed set of row waveform is used with the selection intervals distributed over the entire frame period instead of sequentially to each row. This approach overcomes the decay-time problem that results from the reduced response time of under 50 ms achieved with STN units and is essential to allow a full frame within the 16.7-ms period for a 60-Hz refresh rate. This approach replaces the hardware costs of a fully implemented AMA display with somewhat increased circuit complexity, as shown in Figure 3.22, which shows a fairly simple implementation that cuts the number of AMA drivers in half, by substituting less expensive STN column drivers for the special column pulses. The net result is a display that appears to come fairly close in appearance and performance capability to that achieved with an AMA technique, but at a significantly reduced cost.

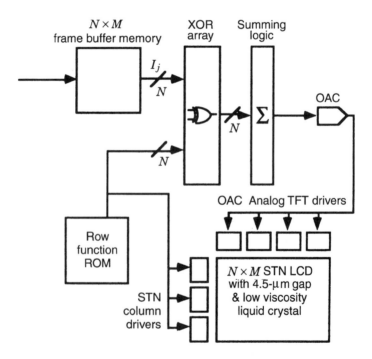

Figure 3.22   Block diagram of a simple hardware implementation of a display system using active addressing. After Scheffer and Nehring [10], by permission of SID.

3.2.7 Field Effect Devices (FEDs)

Field-effect (FE) FPDs have advanced rapidly in the last few years, and have reached the status of viable devices with a number of attractive characteristics. The basic approach is illustrated in Figure 2.34, and is believed to go back over 20 years but is still relevant. The device shown consists of a multiplicity of small field emission cathodes arranged in a matrix, first produced at the Laboratoire d'Electronique et de Technologie de l'Informatique (LETI) in 1986, and the FE cathode technology in turn was developed at Stanford Research Institute (SRI), from which LETI borrowed the concept. The cathodes can emit very high currents when activated by a field across the intersection of a column (gate) and line (cathode). The emitted electrons then activate the phosphor on the anode, and in this respect the technique is similar to that used in a CRT or VFD, except that it is a cold cathode structure. One basic material used for the cathode is silicon, but diamond cathodes have also been fabricated successfully and other materials are possible. There are two types of FEDs: the diode and trode versions as illustrated in Figure 3.23. The luminance (L) for the triode structure is given by [11]

$$L = \eta(V_a)J_{av}(V_{gc})V_a 929 \ (\text{fL}) \tag{3.1}$$

where

$$V_a = \text{anode voltage}$$
$$V_{gc} = \text{gate voltage}$$
$$J_{av} \ (\text{A/cm}^2) = \text{average current density}$$
$$A = \text{emission area (cm)}$$
$$\eta = \text{phosphor efficiency (1/W)}$$
$$N = \text{number of rows}$$

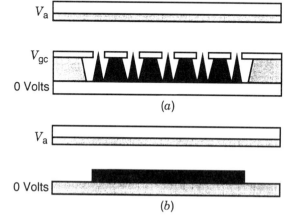

(a)

(b)

Figure 3.23 Basic structure of triode (a) and diode (b) field emission displays. After Xie [11], by permission of IDRC.

TABLE 3.20   FED Parameters

Parameter	Value
Anode voltage (kV)	10
Anode current (mA)	0.45
Luminance (fL)	100
Size (in.)	5.2
Matrix (R × W pixels)	320 × 240
Tips per pixel	384
Phosphor	P-53

Similarly, the contrast ratio is given by

$$CR = 4\exp\frac{(L/V_{gc})}{N} \tag{3.2}$$

Insofar as performance is concerned, luminance values as high as 10,000 fL have been achieved but the more common values are in the range of 40–100 fL. The reported performance parameter values for the 10,000 fL unit are given in Table 3.20. This is an early version, and the reported performance is preliminary. However, it does indicate what can be achieved. In any event, FEDs show considerable promise for the future. Indeed, the favorable reports have led to some newspaper coverage, including a long article in the *New York Times* as long ago as September 28, 1993 so that clearly this technology, in particular that version using diamond emitters, has significant support from the media, which usually leads to increased financial support as well. However, at present this remains only one of several promising approaches and has just reached the stage of providing actual products in any quantity. It remains for the future to improve this situation.

These results appear rather promising, and possibly the FE technology will achieve what none of the others except perhaps LCD has to any significant extent: replace the CRT in almost all applications. The future will tell, but at present it is not possible to predict what will occur.

3.2.8   Other Technologies

The only other technologies used in any significant extent for panels are EM (see driver matrix display in Fig. 3.24) and IN, both of which are found primarily in large-board displays for indoor and outdoor applications. The latter has been used for message displays in outdoors arenas such as sports stadiums. Advertising billboards are another application in which large matrix arrays of standard lightbulbs are used. These are all rather cumbersome and unreliable with high power requirements and are useful only when their high luminance capabilities are important. The displays using EM

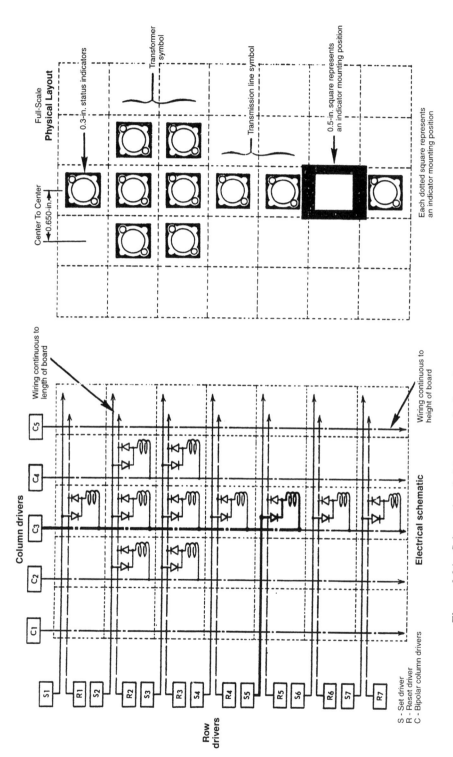

Figure 3.24 Schematic of driver matrix for electromagnetic matrix display. Courtesy of Ferranti-Packard.

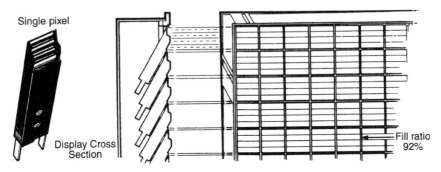

Figure 3.25   Display cross section for Rapidot Display. Courtesy of Densitron.

technology are also limited in their applications but have found some acceptance for large-board message displays, particularly in airports and railroad terminals. There are several implementations as shown in Figures 3.24 and 3.25. The first is described in more detail in Section 2.3.3.9 "FPD Technologies," with parameter values listed in Table 2.12. The second, which is designated "Rapidot," uses similar electromagnetically driven elements in the form of colored vanes driven by miniature solenoids that move in and out of the viewing area. Each pixel can achieve up to six different colors by mixing the primary colors in two positions.

This is the last of the FPD technologies and panel types to be considered here as the other technologies listed in Table 2.1b have not resulted in successful implementations in the form of actual products. This may change in the future as several of those technologies have not been completely forgotten and may result in practical products. However, at present it seems unlikely that this will occur and thus any discussion of their applications does not seem warranted. They should not be completely ignored as history is full of technologies that were rejected at one time and then reappeared when technology improvements made them feasible. But for now, we can only deal with what is available, and therefore, we end with the parameters for Rapidot, given in Table 3.21.

TABLE 3.21   Rapidot Display Parameters

Parameter	Value
Dot size (mm)	24×24
Module pixel format	8×8
Pixel switching speed (ms)	10–70
Switching power (W)	5–10
Operating voltage (V)	10
Colors (no.)	6

3.3   MONITORS

3.3.1   Introduction

Monitors make up a very important product application type for electronic displays. Although at present the bulk of such monitors use the CRT technology, there is a rapid growth in units that use various FPD technologies, in particular EL, plasma, and LC. This growth is due primarily to the proliferation of portable computers such as the notebook, laptop, and palmtop versions, to name the most common types. In addition, with anticipated improvements in performance and reduction in cost for color panels, it may be anticipated that FPD monitors will begin to replace CRT units in desktop units as well. A further extension of applications for monitors is in the new electronic games that use monitor-type displays, and in word processors. These and other major applications for monitors are listed in Table 3.22 along with the technologies most suited for the applications. This is similar in some respects to Table 1.1, as monitors are included in the display systems for all the applications listed in that table.

3.3.2   CRT Monitors

3.3.2.1   Vector Types

The purpose and result of all technologies described previously in relation to CRTs are the CRT monitors (especially for computers and TV receivers), which remain the primary means for viewing the output of a variety of sources, despite strong competition from a variety of PFDs. Therefore, CRT monitors continue to be of considerable interest for those applications that require a flexible display unit capable of high resolution and luminance,

TABLE 3.22   Monitor Applications

Application	Display Requirements	Technologies
Animation	Color, graphics, high resolution	CRT, EL, AMLC
Art	Color, graphics, high resolution	CRT, plotter
Computer graphics	Color, graphics, high resolution	CRT, EL, AMLC, plotter
Desktop publishing	A/N, color, graphics	CRT, AMLC
Electronic games	Color, graphics, high resolution	CRT, AMLC
Geographic information systems	Color, graphics, high resolution,	CRT, plotter
Military	Color, graphics, high resolution	CRT, EL, plasma, AMLC
Multimedia	Color, graphics, high resolution	CRT, AMLC
Television	Color, graphics, high resolution	CRT, AMLC
Virtually reality	Color, graphics, high resolution	CRT, AMLC

rapid response, and excellent visual quality. The CRT has been embedded in a wide variety of units that are termed *monitors*, with a range of performance capabilities. However, the basic system that drives and controls the CRT proper may be represented in its simplest form by the block diagram of a unit that employs magnetic deflection as shown in Figure 3.26. This is also referred to as a *random-deflection* or *vector monitor* in that the X deflection and Y deflection waveforms may be of any type rather than the specific sweep waveforms for X deflection found in oscilloscope and TV monitors. This is discussed in more detail later. An equivalent diagram can be drawn for an electrostatic deflection CRT as is shown in Figure 3.27, and the same considerations apply as for the magnetic deflection unit, except that electrostatic deflection is used. Therefore both may be discussed together.

Briefly, then, the X and Y deflection paths act on the waveform information applied to the two inputs and output amplified waveforms of sufficient amplitude to drive the electron beam to the equivalent locations on the CRT phosphor to produce a visible trace on the CRT faceplate. The magnetic deflection system requires a current and the electrostatic system requires a voltage to achieve these deflections, as has been discussed previously, but otherwise the results are the same insofar as the visual outputs are concerned. However, for an actual visible output to occur it is also necessary to have a video output of sufficient amplitude to permit the beam to pass through the control grid and strike the screen. Then, the duration of the light output is controlled by the video signal, as required, since the video input contains all the visual information from the source, whereas the vertical and

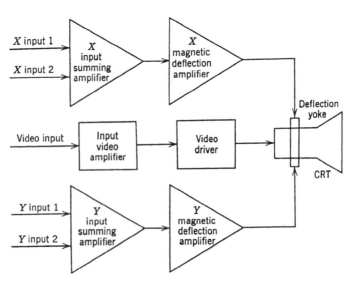

Figure 3.26 Random magnetic deflection monitor flock diagram. After S. Sherr, *Electronic Displays*, 2nd ed., 1993, p. 347. By permission of John Wiley & Sons, Inc.

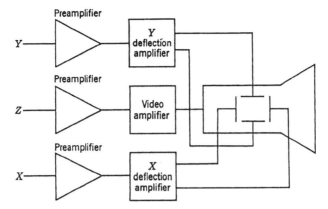

Figure 3.27 Electrostatic deflection CRT monitor block diagram. After S. Sherr, *Electronic Displays*, 2nd ed., 1993, p. 347. By permission of John Wiley & Sons, Inc.

horizontal deflection signals determine the X–Y locations of the beam. In the case of the oscilloscope type of monitor, deflection generators are included so that either the X or the Y deflections may follow a fixed pattern. This is illustrated in Figure 3.28, and several other aspects of monitor operation are included in this version, namely, dynamic focus, sweep failure protection, and high-voltage or ultor supply. These first and last functions are covered in the previous sections in terms of the focus and acceleration electrodes, with sweep failure protections achieved by means of video blanking, and the dynamic focus operation using a circuit whereby the focus voltage control is automatically set to optimize the beam for different luminance settings.

3.3.2.2 Raster TV

Although vector monitors still serve a function in oscilloscopes, their more general applications have been taken over by raster types, of which the TV monitor is the prototype. The block schematic for the basic unit is essentially the same as the one shown in Figure 3.26 for the magnetic deflection vector unit, with the deflection generators designed to operate in the TV mode, that is, with refresh rates of 525–1280 lines for the horizontal or X deflection and 30–70 Hz for the vertical. This is discussed in more detail later in the section dealing with raster systems. Otherwise, the same considerations hold for the monochromatic versions as for the equivalent vector monitors, and no further discussion is needed. However, color monitors introduce a whole new set of requirements, and they are covered in some detail next.

Color monitors use either the shadow-mask or Trinitron versions of color CRTs, and the general circuit configurations are the same for both, so a single diagram will do for this discussion. This diagram, shown in Figure 3.29, differs from the monochromatic unit in that the color CRT is a three-gun

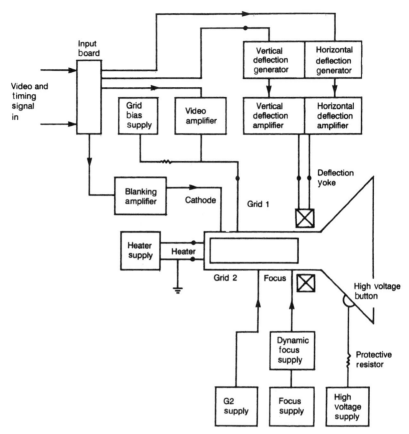

Figure 3.28   Block diagram for oscilloscope CRT monitor. After Santelmann and Mason [12a], by permission of SID.

tube, and there are three color inputs to the video amplifiers, as well as the horizontal and vertical synchronizing signals that drive the two deflection generators. In addition, there may be a number of specialized circuits such as graphics boards, which may be any of those lited in Table 3.23. These go from low resolution through medium and high resolution to very high resolution, with further development possible as the demand for higher resolution and more colors continues to grow.

Other circuits that are usually included are the sweep failure protection and dynamic focus ones previously referred to in the section on vector monitors. The first is important to protect against phosphor burnout, and the second allows for improved focus without the need for manual adjustment of the focus control. Other capabilities that might be included are circuits for dynamic convergence to minimize color distortion and pincushion correction and correction of flat-face distortion to achieve low geometric distortion.

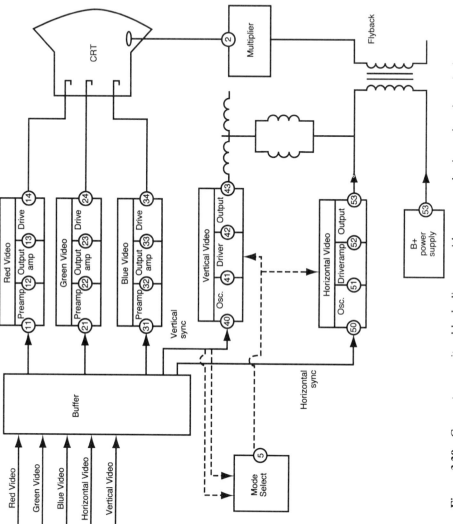

Figure 3.29 Computer monitor block diagram with separate horizontal and vertical sync lines. Courtesy of Sencore, Inc.

TABLE 3.23   Graphic Boards

Standard[a]	Resolution	Colors No.
EGA	640 × 350	256
VGA	640 × 480	256
SVGA	800 × 600	256
	1024 × 768	16
8514a	1024 × 768	256
XGA	1024 × 768	256
	640 × 480	65,536
TIGA	4000 × 4000	256

[a]EGA, enhanced graphics adapter; VGA, video graphics array; SVGA, super-VGA; XGA; 8514a IBM XGA; TIGA, Texas Instruments Graphics Architecture.

The parameter values for a color monitor to be used in a full-graphics system are expanded by the use of the most advanced graphics boards, and these boards then fall in the range of those shown in Table 3.24. They can be expanded further as additional types with more capabilities become available and replace those presently in use. Thus resolution of 1280 × 1024 and higher may become the new standards. These are the most common values, but in some cases they may be exceeded for special-purpose applications. Therefore, the performance range for vector monitors is well covered by these parameter ranges. However, to fully understand the significance of the parameters, it is advisable to discuss the meaning of the parameter types in terms of the raster approach (see Fig. 3.30). It differs from vector approach in that the X and Y deflection signals are fixed by whatever the standard is, for example, 60 Hz vertical and 31.5 kHz horizontal, corresponding to 60-Hz refresh rate and 525 scanning lines. This is representative, although the exact NTSC (National Television System Committee) values are 59.94 Hz and 15.73 kHz, respectively, chosen to be compatible with the line frequency divided by 2 for interlacing to reduce video bandwidth requirements, as shown in Figure

TABLE 3.24   Graphics Color Monitor Parameters

Parameter	Full-Graphics Value	Range
Resolution (TV lines)	1280 × 1024	640 × 480–4000 × 4000
Color palette (maximum number of colors)	16.7 million	16–4096
Active diameter (in.)	27	19–35
Refresh rate (Hz)	60(N)[b]	30(I)[a]–72(N)[b]
Video bandwidth (MHz)	20	3.5–100

[a]Interlaced.
[b]Noninterlaced.

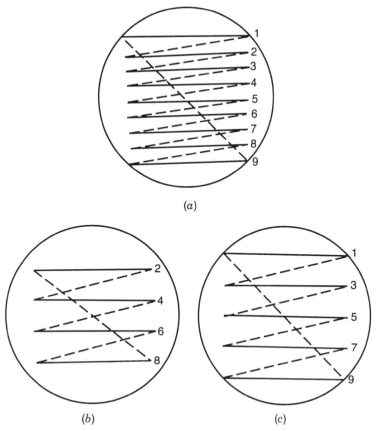

(a)

(b)   (c)

Figure 3.30 Raster scan patterns: (a) noninterlaced; (b) interlaced even lines; (c) interlaced odd lines. Reprinted from Sherr [12b], permission of John Wiley & Sons, Inc.

3.30, and the number of scanning lines. This leads to a video bandwidth requirement of about 3.56 MHz in order to resolve the horizontal data. Similarly, the PAL and SECAM versions used in Europe and elsewhere have 50 Hz divided by 2 and 625 scanning lines, leading to the equivalent numbers for these systems. All of these will change drastically when high-definition TV (HDTV) is broadcast at a resolution of something in the vicinity of 1280 × 1024 pixels and a noninterlaced vertical scan that might require video bandwidths approaching 15–20 MHz for best performance.

These TV requirements have a significant impact on the performance parameters for CRT monitors, and the computer applications can add to these, bringing the vertical scanning rate up to over 70 Hz to avoid flicker, and the resolution to 2000 × 2000 or higher to achieve photographic quality. In these cases, the video bandwidth may exceed 100 MHz, leading to stringent design requirements to attain this type of performance. Units are available that meet these requirements, and it can be assumed that they will be improved and the cost reduced sufficiently to make them available for a large number of applications that presently are limited to the performance capabilities of standard TV monitors for reasons of cost and/or availability.

3.3.3 FPD Monitors

3.3.3.1 Electroluminescence (EL)

Monochromatic EL monitors have been available for some time but have found limited application. However, with the advent of color versions, the applications for these units have expanded, and it may be anticipated that they will find a market, in particular for those applications that benefit from the better form factor and lower voltage requirements for EL FPD than for CRT monitors. Both TFEL and dc EL monitors are available, and the block diagram of one TFEL unit is shown in Figure 3.31. This is an earlier monochromatic unit, and there have been several approaches to fabricating color units. One of the most successful structures to date for a color unit, shown in Figure 3.32, employs a hybrid construction of stacked phosphor layers, with the top layer containing patterned red and green phosphors and the bottom layer, a patterned blue layer. The red color comes from a filtered $ZnS \cdot Mn$ phosphor; the green, from a $ZnS \cdot Tb$ phosphor film; and the blue, from a $CaGa_2S_4 : Ce$ phosphor, which has a blue-and-green emission pattern.

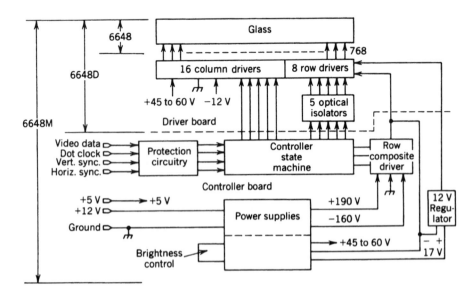

Figure 3.31 Block diagram of TFEL monitor. Courtesy of Planar Systems, Inc.

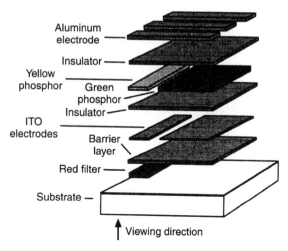

Figure 3.32 Structure of multicolor EL monitor. After King [12c], by permission of SID.

Other approaches have used a white phosphor with filters in the structure shown in Figure 3.10*a*. Development is continuing on these and other approaches, and it may be anticipated that several different types of color EL monitors will be available. The performance parameters for the unit shown in Figure 3.32 are given in Table 3.25, and it approximates VGA performance. The panel color range is compared with that for a CRT monitor in Figure 3.33.

It is apparent from these values that EL monitors still have a considerable way to go before they offer performance capabilities that are truly comparable with that available from CRT units. However, the improvements that

TABLE 3.25   EL Monitor Parameters

Parameter	Value
Color pixel matrix	640 × 480
Number of colors	16
Color pixel pitch	0.31 mm
Subpixel size:	
Red	0.060 × 0.210 mm
Green	0.170 × 0.210 mm
Blue	0.210 × 0.210 mm
RGB luminance:	
Red	20 fL
Green	30 fL
Blue	3 fL

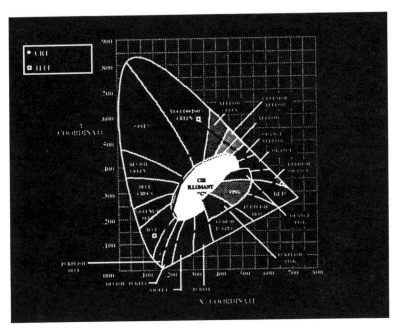

Figure 3.33 Color range for dual-substrate full-color TFEL compared with color range for color CRT. After King [12c], by permission of SID.

have been achieved are certainly impressive, and further development should lead to viable units before too long. Therefore, it is advisable to consider the EL types as possible replacements for the CRT units in the relatively near future. The values were measured at 60 Hz and 40 V phosphor drive. The blue output is increased in the panel by means of making the blue phosphor pixel size to approximate the size of the sum of the red and green pixel. This is made possible by the dual-substrate structure, which also allows the blue phosphor to be driven at a higher refresh rate, also increasing the blue output luminance. Thus, this panel overcomes, at least to some extent, the limitations in blue output, and achieves the results listed in Table 3.25, which are not spectacular but do offer future promise.

3.3.3.2 *Plasma*

Plasma panels have been most notable among FPDs in their large size and good luminance capabilities. However, a feasible color approach has been developed for ac units only recently, and such units are available in a monitor format from several sources. Other monitor units that are available use monochromatic display panels with the customary neon-orange color. One structure used for such a panel is shown in Figure 3.34, and the block diagram for a monitor using the panel as the display is shown in Figure 3.35. This monitor contains an interface unit that provides the necessary synchro-

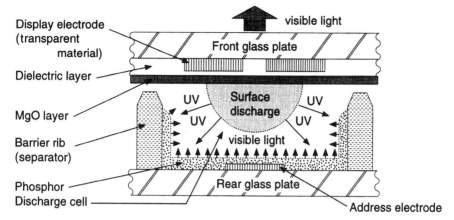

Display electrode (transparent material)

Dielectric layer

MgO layer

Barrier rib (separator)

Phosphor

Discharge cell

visible light

Front glass plate

UV   Surface discharge   UV

UV   UV

visible light

Rear glass plate

Address electrode

Figure 3.34 Physical construction and operation of Fujitsu's ac-type plasma displays. Courtesy of Fujitsu.

nizing and clock signals, and a console unit that contains the various control elements. In this respect it is quite similar to the standard CRT monitors. The matrix format of the display panel makes it visually somewhat different in its appearance. However, this potential deficiency is overcome by the high-resolution capability of the panel. In addition, the unit is available in a 16-in.-diameter screen size, so that it can compete with the equivalent CRT monitors in its general capabilities, except for the lack of color. However, this lack is overcome by another model with a matrix panel resolution of 852($\times$3) $\times$ 480, accompanied by a larger screen diagonal size of 42 in., over 16 million colors, and the ability to accept either RGB or NTSC signals through the interface board. The physical appearance of the latter unit is shown in Figure 3.36, and the performance parameter values for the former, given in Table 3.26, are representative of these types of plasma monitors. Another color panel unit has a higher matrix resolution (1024 $\times$ 768) at a diagonal size of 30 in., so higher resolution monitors may be expected. This type of performance, when coupled with the emerging color capabilities, should make ac plasma monitors quite competitive with the CRT-based units. In addition, insofar as dc plasma units are concerned, they also exhibit impressive advances in color developments; the latest are a 40-in. diagonal panel for HDTV and an 18-in. diagonal panel for NTSC, both with a full-color capability. However, these units are not available as commercial products in monitor format, but these and other dc plasma color monitors are expected to enter the market at some future date.

3.3.3.3 Liquid Crystal (LC)

The most successful monitor types using FPD technologies are those based on the active-matrix LCD (AMLCD) approach, in particular for color dis-

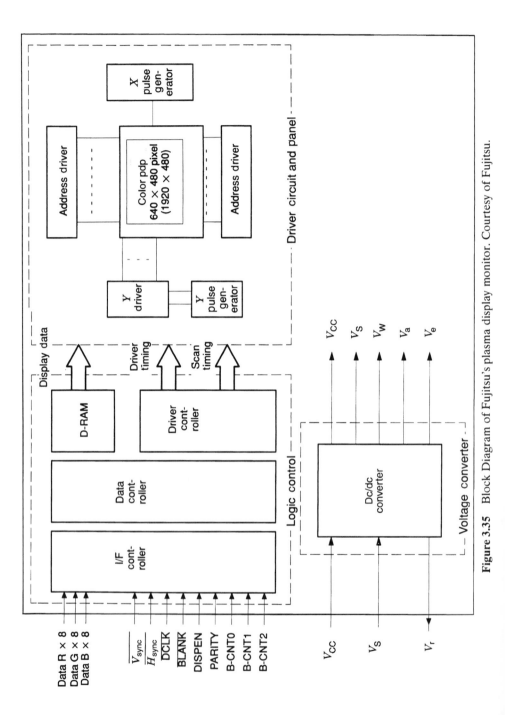

Figure 3.35 Block Diagram of Fujitsu's plasma display monitor. Courtesy of Fujitsu.

Figure 3.36 Fujitsu large-screen full-color plasma display. Courtesy of Fujitsu.

TABLE 3.26 Plasma Monochromatic Monitor Parameters

Parameter	Value
Display matrix (H × V[a])	1280 × 1024
Dot pitch (H × V in.)	0.010 × 0.010
Dot size (diameter in.)	0.006
Display area ($H \times W$ in.)	12.6 × 10.08
Luminance (cd/m^2)	7
Contrast ratio (at 500 lux)	20: 1

[a] Horizontal × vertical.

plays—this despite certain advantages found in the plasma units in terms of maximum size, and for EL in small, high-resolution displays. One reason for this advantage has been the proliferation of portable computers that require the form factor provided by FPDs and the ability for color AMLCD panels to meet all the requirements for such computers. However, the majority of these displays, although fulfilling the entire monitor function satisfactorily, are not produced as independent units, but are connected directly to the computer assembly. Therefore, it is perhaps not quite correct to classify them as monitors, but rather as panel display units. However, they are so close to independent monitors that it appears appropriate to discuss them in this section along with the limited number of examples of true, standalone monitors using the LC technologies.

The basic structure of a compensated supertwist LCD using polymer film to achieve a black-on-white display is shown in Figure 3.19, and Figure 3.21 demonstrates the effect of a similar structure using color polarizers contained in a white display. Similarly, parameter values for the monochromatic and color versions of LC panels are presented in Tables 3.18 and 3.19, respectively. A somewhat more detailed design for a subtractive color approach that illustrates how the three basic colors for an RGB system are generated by using color filters, and assuming a white light source, is shown in Figure 3.37, and demonstrates the complexity of the design. Three different polarizers are required for the three basic colors, which further complicates the structure, but the results are the closest to color CRT images that any of the technolo-

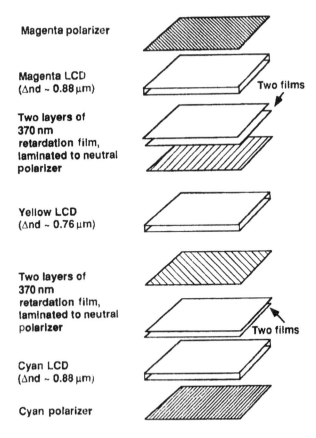

Magenta polarizer

Magenta LCD
(Δnd ~ 0.88 μm)

Two films

Two layers of
370 nm
retardation film,
laminated to neutral
polarizer

Yellow LCD
(Δnd ~ 0.76 μm)

Two layers of
370 nm
retardation film,
laminated to neutral
polarizer

Two films

Cyan LCD
(Δnd ~ 0.88 μm)

Cyan polarizer

Figure 3.37 Construction of triple supertwisted nematic (TSTN) system. After Conner [13], by permission of SID.

TABLE 3.27   LCD FPD Monitor Parameters

Parameter	Color Model	B & W Model
Overall dimensions (mm)	$315W \times 270H \times 158D$	Same
Viewing area (mm)	$197W \times 147H$	$215W \times 163H$
Diagonal (mm)	207.9 (8.2 in.)	269.2 (10.4 in.)
Display format (matrix)	$640W \times 3 \times 480H$	$640W \times 480H$
Dot pitch (mm)	0.10×0.30	0.32×0.32
LCD technology	STN dual-scan passive matrix	B & W STN
Colors (no.)	256	2
Gray scale	None	32
Power (W)	8	8

gies have achieved. However, another approach, termed *dual-scan passive matrix*, which differs from the passive matrix addressing described in Section 3.2.6.2 in that it uses a double parallel scan to improve response time, has also managed to produce units with acceptable color. In addition, black-and-white (B & W) models using the more standard STN matrix technology are also available as FPD monitors. Representative operating parameter for both types are given in Table 3.27. These are noteworthy performance capabilities, and the units may be used for many applications with the advantages of low radiation, low power consumption, and compact size.

However, good as these passive matrix monitors are, the color performance is still inferior to the AMLCD designs that have resulted in very acceptable color FPD monitors, with a number of units available that achieve VGA and SVGA performance. One approach to the structure of the panel is shown in Figure 3.38, and the parameter values for three representative color units are given in Table 3.28. These values compete with those of CRT monitors except for display size and SVGA resolution. However, panels are

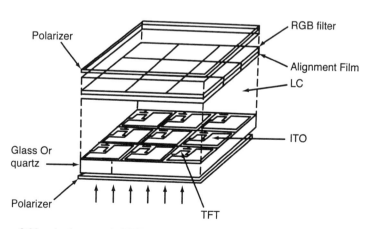

Figure 3.38   Active matrix TFT LCD. After Schlam [14], by permission of SID.

TABLE 3.28   AMLCD Color Monitor Parameters

Parameter	Value 1	Value 2	Value 3
Screen size (diagonal in.)	10.4	8.6	5.6
Colors (nos., million)	16	16.7	16.7
Pixels (RGB trio)	307,200	145,920	56,160
Horizontal resolution (TV lines)	—	370 +	250 +
Pixel resolution (H × V)	640 × 480	640 × 228	480 × 117
Computer input	VGA	No	No

available with 9.4-in.-diameter screens and 800 × 600 resolution, so that in at least one performance factor the AMLCD color monitors meet CRT performance without the associated radiation, power requirements, and size, although the cost is significantly higher. It does not appear that larger sizes can be readily achieved with these panels and monitors, but the answer to that limitation may be found in the LCD projection panels discussed in Section 3.5.3.2. In any event, this completes the discussion of LCD monitors.

3.4   HARD COPY

3.4.1   Introduction

Although electronic displays are supposed to eliminate the need for hard copy, hard-copy devices remain in extensive use. These devices include all types of producing units that cause a permanent visual record to be generated. The continuing proliferation of these devices and their output make them an important part of the total display-generating group, and although it is more precise to term them *electrical* rather than *electronic* displays, they have enough in common with the latter type to be included in this volume. In addition, they are almost always part of electronic display systems, and their similarity to the non-hard-copy displays discussed previously is emphasized by their description as hard-copy displays versus the designation of soft-copy displays accorded to the other group. For convenience of reference, such devices are listed in Table 3.29, with some information about the technologies used, although the discussions in this chapter are devoted primarily to the technologies, with the devices covered in Chapter 3. Essentially all the major technologies are included in Table 3.29, although a few still in the early stages of development may be missing.

One unit missing from this list is the typewriter in any of its forms, but this device is so little used when electronic displays are involved and is so well known that it does not warrant any further discussion. Another is the letter-quality printer using a fixed typeface. Then there are also the impact printers that use fixed typefaces primarily for large-scale printing and do not fall within the purview of this volume. The technical discussions found in the

TABLE 3.29   Hard-Copy Devices

Device Name	Technology	Types
Plotter	Pen	Drum, flatbed
	Electrostatic	Single-head, four-head
Printer–plotter	Inkjet	Continuous, piezoelectric
	Electrophotographic	Laser
	Thermal	Direct transfer, dye sublimation
	Dot matrix	9-pin, 24-pin
Film	CRT	Electrostatic, fiberoptic

next sections cover all the major approaches to hard copy production in electronic display systems.

3.4.2   Technology

3.4.2.1   Pen Plotter
The pen-plotter technology is one of the oldest types used in conjunction with the soft-copy displays found in electronic display systems. Pen-plotters are finding somewhat less acceptance with the proliferation of other technologies, but remain as some of the most reliable devices capable of generating high-quality images over a wide range of sizes and costs. It should be noted that pen-plotters are defined by the designation of *pen*, which signifies that some type of pen is used to create the image in each version, although the exact format and structure of the material that receives the ink may differ.

There are two approaches to the application of the basic technology: *flatbed* and *drum*. Examples of the flatbed design are pictured in Fig. 3.39, and differ primarily in the size of the hard-copy surface. In this type of embodiment, the hard-copy material—usually paper, but other materials are also possible—is held constant while the pen moves and writes on the paper in an *X–Y* analog mode. A number of pens, containing different-color inks, are available, and each can be separately selected so that as many as eight different colors may be used on the same drawing. The result is a high-quality drawing, with resolution as high as 0.0002 in., and accuracy between 0.1 and 0.3% of the amount moved. These values are shown in Table 3.30, where they are compared with those for drum plotters. The forms taken by drum plotters is shown schematically in Figure 3.40. They consist of a pen that moves horizontally across the paper, which is caused to move vertically by means of a grit wheel–pinch roller combination as shown in Figure 3.41. The grit wheel consists of a drum that is coated with tiny particles of aluminum oxide or similar material and a rubber pinch roller that presses the paper against the drum. The ink in the pen is transferred to the paper while the pen and paper are both moving in orthogonal directions. As in the case of the

DPX-4600A
ANSI-E/ISO-A0 SIZE

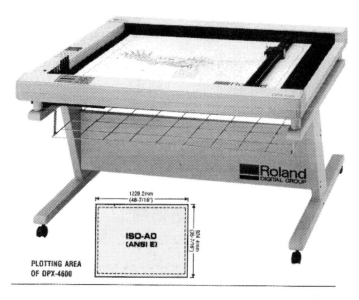

DPX-3700A
ANSI-E/ISO-A1 SIZE

DPX-2500A
ANSI-E/ISO-A2 SIZE

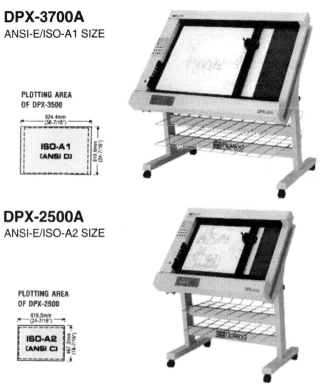

Figure 3.39   Flatbed plotters. Courtesy of Roland Digital Group.

TABLE 3.30   Plotter Parameters

Parameter	Flatbed	Drum
Media size (in.)	8 × 11–36 × 48	11 × 17–36 × 48
Pens (no.)	1–8	1–14
Resolution (in.)	0.0002–0.0035	0.00006–0.005
Accuracy (% of move)	0.1–0.3	0.01–0.5
Repatability (in.)	0.001–0.005	0.0004–0.005

flatbed unit, the drum plotter may have as many as eight separate pens that can be selected. A full set of performance parameters for flatbed and drum plotters is given in Table 3.30.

3.4.2.2   *Electrostatic Plotter*

The electrostatic plotter has much in common with the drum plotter in that the paper moves orthogonally to the writing source, but it differs significantly in the manner in which the visible traces are generated by the writing head and transferred to the hard-copy material. The writing is done by applying a charge to the writing surface by means of a writing head, and then transferring liquid toners to the charged surface by moving the hard-copy material past the liquid toner modules, as shown in Figure 3.42. There may be as many as four toners, containing liquids that are colored cyan, magenta, yellow, and black so that a full color range is available. This capability for full color is relatively recent, and monochrome plotters are also available at lower cost and higher speed if only one writing head is used; so thus four writing sequences are required with the four toners. This may be overcome in the color plotter by having four writing heads so that it may perform in parallel, but at increased cost. However, the four-head unit is most popular, and has performance capabilities that match those of the other plotters covered previously, and exceed them in speed capability because the writing heads can produce multiple lines simultaneously rather than one at a time.

3.4.2.3   *Inkjet Printer–Plotter*

This form of hard-copy device uses a technique that differs considerably from that used by the pen plotter, but has some similarities to the technology employed by the electrostatic plotter, in that it is nonimpact in the sense that the only physical contact between the writing surface and the source is by means of a stream of writing fluid. Another similarity to the electrostatic plotter is that the writing fluid consists of drops of ink that are electrostatically charged, as is the case for the toners. The ink is emitted in a continuous flow or a "drop on demand" basis, and Figure 3.43 represents a design for the former, where the ink is emitted in a stream of dots that are charged to a level determined by the amount of deflection required for the final location of the dot. Thus, the deflection technique is quite similar to that used for the

MODEL   **OVERVIEW**

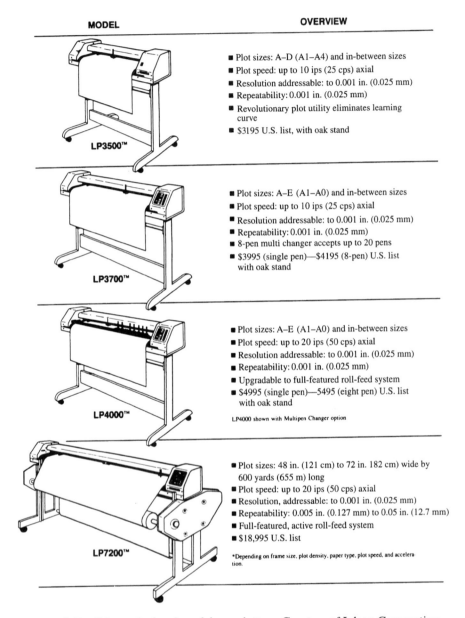

LP3500™

- Plot sizes: A–D (A1–A4) and in-between sizes
- Plot speed: up to 10 ips (25 cps) axial
- Resolution addressable: to 0.001 in. (0.025 mm)
- Repeatability: 0.001 in. (0.025 mm)
- Revolutionary plot utility eliminates learning curve
- $3195 U.S. list, with oak stand

LP3700™

- Plot sizes: A–E (A1–A0) and in-between sizes
- Plot speed: up to 10 ips (25 cps) axial
- Resolution addressable: to 0.001 in. (0.025 mm)
- Repeatability: 0.001 in. (0.025 mm)
- 8-pen multi changer accepts up to 20 pens
- $3995 (single pen)—$4195 (8-pen) U.S. list with oak stand

LP4000™

- Plot sizes: A–E (A1–A0) and in-between sizes
- Plot speed: up to 20 ips (50 cps) axial
- Resolution addressable: to 0.001 in. (0.025 mm)
- Repeatability: 0.001 in. (0.025 mm)
- Upgradable to full-featured roll-feed system
- $4995 (single pen)—5495 (eight pen) U.S. list with oak stand

LP4000 shown with Multipen Changer option

LP7200™

- Plot sizes: 48 in. (121 cm) to 72 in. 182 cm) wide by 600 yards (655 m) long
- Plot speed: up to 20 ips (50 cps) axial
- Resolution, addressable: to 0.001 in. (0.025 mm)
- Repeatability: 0.005 in. (0.127 mm) to 0.05 in. (12.7 mm)
- Full-featured, active roll-feed system
- $18,995 U.S. list

*Depending on frame size, plot density, paper type, plot speed, and acceleration.

Figure 3.40   Schematic drawing of drum plotters. Courtesy of Iolene Corporation.

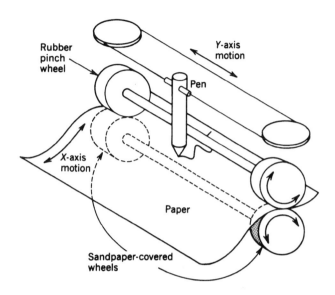

Figure 3.41 Drawing of grit wheel paper drive mechanism. After Patterson and Lynch [15]. Copyright © 1981 Hewlett-Packard Company. Reproduced with permission.

electrostatic deflection CRT in that electrostatic deflection plates are used, as shown in Figure 3.43, and the amount of deflection (D) is given by

$$D = \frac{qe}{2mV^2L(Z - L)} \qquad (3.3)$$

where q = charge on drop
   e = deflection field

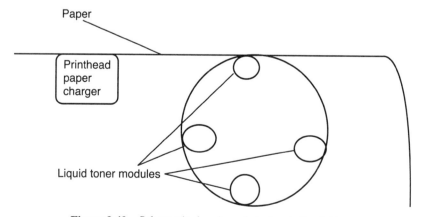

Figure 3.42   Schematic drawing of electrostatic plotter.

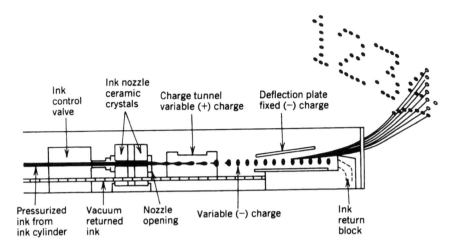

Figure 3.43 Schematic drawing of continuous-stream inkjet printhead. Courtesy of Videojet Systems International, Inc.

L = deflection length
m = mass of drop
V = drop velocity
Z = distance from deflection plates entrance to medium

The drops are created by means of piezoelectric action on the part of the ceramic crystals, and the jet is formed by hydrostatic action that drives the ink through the small opening in the nozzle. The net result of this sequence of actions is the deposit of the inkdrops on the writing surface in the locations determined by the input charge data. Other forms of continuous-inkjet systems exist, but the one shown in Figure 3.43 is sufficient to demonstrate the basic principles.

The alternate form of inkjet printer that is termed "drop-on-demand" (DOD) is illustrated in Figure 3.44. The dot charging and deflection is accomplished in a similar fashion as for the continuous system, but the dots are created by means of a piezoelectric crystal that generates a pressure wave to force the drops through the nozzle only when called for. Another approach to DOD is the use of thermal techniques to vaporize a small amount of ink and eject drops through a very small orifice. This latter method, developed by Canon and Hewlett-Packard (HP), has been termed "paint jet" by the latter organization. It should be noted that the thermal approach seems to dominate, using some version of the Canon engine, and also that color is achieved by using multiple-color inks and nozzles and high-quality color images can be produced by either version.

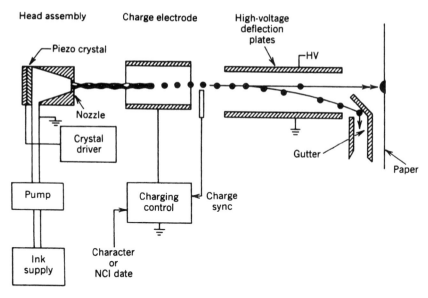

Figure 3.44 Schematic diagram of Hertz continuous-jet printhead with media on drum. After Mills [16], by permission of SID.

3.4.2.4 Laser Printer–Plotter

The laser printer–plotter is the only presently available unit that uses electrophotographic technology, but it has reached the point in its development and commercial availability that makes it an important addition to the hard-copy group. The technology shows great similarities to the other important example of this technology: the Xerox-type paper copiers. The process uses an electrostatic charge to deposit toner on the print medium, and the laser discharges areas selectively on a photoconductive sheet. The discharge pattern is determined by the sequence of locations of the laser beam, where these locations are determined by the controlled writing sequence. The toner is then transferred to the charged areas in a manner similar to that used for the electrostatic plotter, and fused by melting. Color is achieved by repeating this sequence once for each primary color, and full-color systems are now available. The resolution is determined by the beam size of the laser and can be very high, with 600 × 600 dpi available in monochrome, and 300 × 300 dpi (dots per inch) in both monochrome and color, although one vendor (Xerox) offers 1200 × 300 dpi, and another (QMS) provides 600 × 600 dpi in color, both at the cost of additional RAM (random-access memory). Printing speeds range from 12 ppm (pixels per minute) for 300-dpi monochromatic mode to 2 ppm in full color. These many sources indicate that the color laser units have come of age, although the high prices will keep them from wide acceptance for some time.

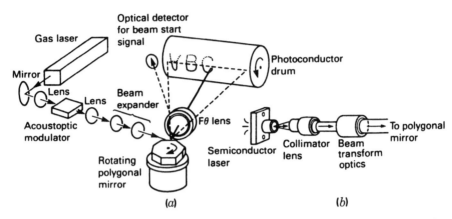

(a)   (b)

Figure 3.45 Two versions of a laser printhead: (*a*) a typical gas laser printhead; (*b*) a typical diode laser printhead. After Jaffe and Burland [17, p. 235], by permission of Academic Press.

There are several types of laser printhead; two of the most common ones are shown in Figure 3.45. These represent the use of the gas laser in one case and the diode laser in the other. The diode laser offers some simplification and is used by a larger number of manufacturers. In addition, some printers that employ the electrophotographic technology use LEDs and LCD shutters in units that are also termed *laser printers*, and are identical in appearance from the outside.

3.4.2.5   Thermal Printer–Plotters
Although thermal printing is a fairly old technology, going back to the 1960s, when the procedure was to use heat-sensitive paper as the hard-copy surface, with the printing done by an array of heater pins, as illustrated in Figure 3.46. This method is still in use, with the heat-sensitive paper containing a heat-sensitive coating of dye-forming chemicals that are colorless at room temperature but acquire color when heated. Thermal technology is also used in plotters that use raster scanning techniques similar to those used for electrostatic plotters. The results are fairly acceptable, but the quality is not as good as for the electrostatic and laser plates, and the paper cost is higher.

Although the direct thermal printer–plotter has some good features, the need for special paper is a drawback, and this technology is being replaced by thermal transfer printing on plain paper, as shown in Figure 3.47. The main difference from direct thermal technology is in the addition of a ribbon that has a low-melting-point, waxlike ink coated on a thin paper of polyester film carrier layer. A resistive ribbon may also be used, that is heated by current flowing through from the printhead. Finally, there is the dye-sublimation thermal transfer where the dye is refined before being transferred, so that it

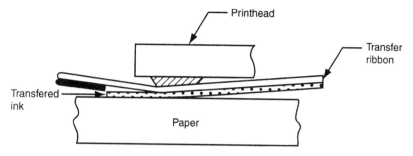

Figure 3.46   Schematic drawing illustrating thermal transfer printing. After Sahni [18], by permission of SID.

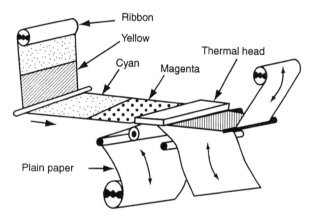

Figure 3.47   Color thermal transfer printing. After Sahni [18], by permission of SID.

can be diffused directly into the hard-copy medium, resulting in what is possibly the best-appearing hard copy. In any event, the various printer–plotters using some one of the thermal hard-copy technology all offer viable approaches to hard-copy devices. The technologies allow 300×300 and 300×600 resolutions, with three or four color ribbons, and excellent color rendition, albeit at a somewhat higher price than inkjet.

3.4.2.6   Dot-Matrix Printer–Plotter
The final hard-copy printing technology to be discussed in this brief summary is the impact dot-matrix approach. This for many years was the preferred—if not quite the only—way to achieve acceptable results in printers, if not necessarily in plotters. It had some deficiencies in appearance as long as the 9-pin printhead was in wide use, but with the general transition to 24-pin units, the quality is much improved, and the claim is that letter quality can be achieved. The basic printhead consists of a matrix of wires (see Fig. 3.48)

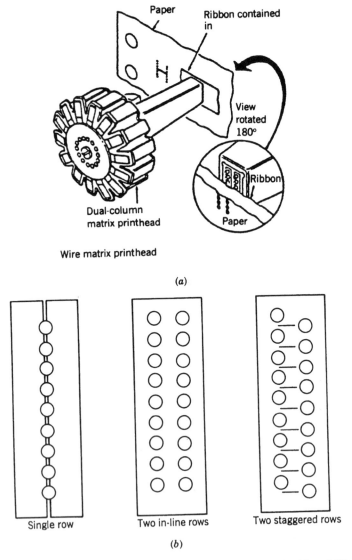

Figure 3.48 (*a*) Schematic drawing of wire matrix printhead. After Mills [16] by permission of SID. (*b*) Types of print guides used for wire matrix printing. After Williams [19, p. 178], by permission of Academic Press.

located alongside the paper, with a ribbon between it and the paper. The pins are selected to produce the desired character or graphics image for each position, and the entire printhead is moved along the paper horizontally from one print position to the next until the full line is printed, at which point the paper is made to move up one line and the printhead returns to its original starting position. High-quality character print is achieved by using matrices as high as 35 × 48, although matrices as low as 7 × 9 may be used for lower quality and still be quite legible. Graphics with resolutions as high as 360 × 360 are possible with some units, and color is attainable by using multiple ribbons. The technology leads to somewhat noisy printers, and the quality is less than that possible with the other technologies, but the low cost makes this an attractive technology that is still in wide use.

3.4.2.7 *Film-Based Technology*

The last technology that is used to any significant extent for hard-copy output is that based on the use of photographic film. This is a fully electronic technique except for the final photographic development process, as opposed to the combination of electronic and mechanical procedures used in the other hard-copy technologies. The basic approach is to image the desired output on the face of an electronic display such as a CRT, and then photograph this output image with an appropriate camera associated with a film transport that can move the film to the next frame when the image has been exposed. This sequence requires a complete electronic system, as illustrated in the block diagram in Figure 3.49, where the CRT is shown as the visual device, and is the one most commonly used in film-based hard-copy systems, although other visual devices are also feasible. The electronic portion operates like one of the monitor-type output systems discussed previously, and the camera may be one of several types that produce movies, stills, or transparencies. There is also an appropriate lensing system between the image and the camera, and slide production is probably the most important application for this technology. It should be noted that some systems achieve color by using color monitors, but the highest-quality systems have adopted a color wheel approach in which color is obtained by rotating a color wheel in front of a monochromatic, high-resolution monitor.

3.5 LARGE-SCREEN SYSTEMS

3.5.1 Introduction

Large-screen systems constitute the last category of output devices and systems covered in this chapter, and make up a frequently used supplement to or replacement for the monitor group of display products. There are a number of different forms in which large-screen systems in general and projection devices or large-board displays in particular may be found, with a

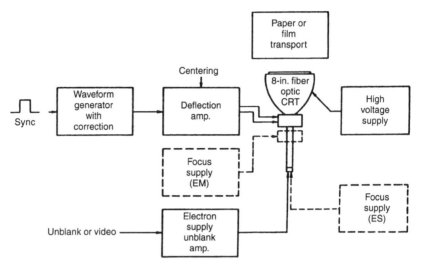

Figure 3.49 Block diagram for CRT recorder or scanner. After Wurtz [20, p. 178], by permission of Academic Press.

concomitant set of choices for the systems of which the projectors and large boards per se are the final output element where the resultant imagery may be viewed. The various types and technologies that are available are listed in Table 3.31. These are the major types in use, although the direct-projection CRT, Schlieren light-valve, and direct-projection LCD systems are the most common. However, the others are not completely forgotten and are treated to at least a limited extent in the following sections.

3.5.2 CRT Projection

3.5.2.1 Schmidt
Schmidt projection is probably the oldest of the commonly used CRT projection system, and was the technology of choice for home and theater

TABLE 3.31   Large-Screen Systems

Type	Technology	Output	Optics
Projection	CRT	Screen	Schmidt
Projection	CRT	Screen	Rear mirror
Projection	Light-valve	Screen	Schlieren
Projection	CRT	Screen	Scheimpflug
Projection	LCD light-valve	Screen	LCD
Projection	Liquid crystal	Screen	Direct
Board	Magnetic	Reflective	Matrix
Board	Incandescent	Emissive	Matrix

projection before the advent of the Schlieren and LCD light-valve systems. Its major advantage was relatively low cost, but it lacked adequate luminance and resolution to be completely satisfactory for many commercial applications. It is built around the special CRT types that are designed to emit at high luminance levels directly into a spherical mirror, as shown in Figure 3.50. The corrective lens compensates for the spherical aberrations created by the mirror, and this is a high-efficiency optical system. It is theoretically possible to use a standard color CRT such as the shadow-mask or Trinitron types, but the output from these types is much too low to be acceptable, and color systems require three separate CRTs with color filters to achieve adequate output levels in full-color projection systems. This leads to some problems in aligning the optics, which is overcome by the direct-projection LCD system discussed later. This type of system has the advantage of reduced throw distance over front-projection systems.

3.5.2.2 Rear Mirror

The need for a correction lens is eliminated in rear-projection systems with CRTs as the image source by the use of one or two mirrors as shown in Figure 3.51, with the mirrors not part of the objective lens. For this type of system, the most important optical parameter is the *total conjugate length*

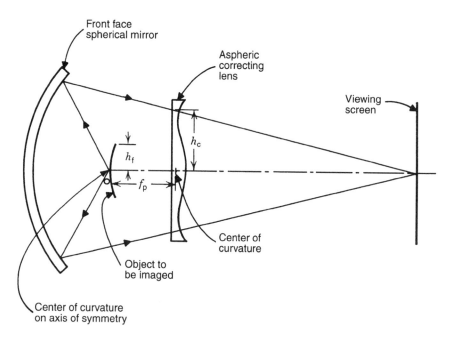

Figure 3.50 Schmidt projection system. After Pender and McIlwain [21], by permission of John Wiley & Sons, Inc.

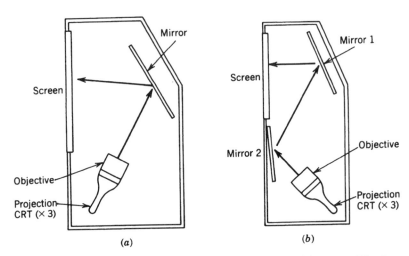

Figure 3.51 Three projection CRTs are combined with one (*a*) or two (*b*) mirrors to reflect image onto screen in rear projection system. After Gibilini and House [22]. Copyright © Pennwell Publishing Co.

(TCL), which is defined as the distance between an object and the image of that object produced by the length. It is desirable to keep that number as small as possible in order to minimize the size of the cabinet. The focal length of the projection lens is given by

$$f = \frac{m(\text{TCL})}{(m + 1)^2} \tag{3.4}$$

where f = focal length
 m = active display area/usable screen area

This shows that the shorter the TCL, the shorter the required focal length of the lens and the required distance between the lens and the screen. This is a desirable condition as it reduces the cabinet size as required for acceptable form factor. It should be noted that three projection CRTs are needed for a full-color CRT system.

3.5.2.3 Front Projection
Front-projection CRT systems include a modified version of the Schmidt system, and a somewhat simpler configuration, shown in Figure 3.52. These may be termed *reflective* and *transmission* systems, respectively. It should be noted that the requirement for a short TCL does not necessarily hold. However, mirrors cannot easily be used to make the optical axis of the lens perpendicular to the screen as is done for rear-projection systems, and the

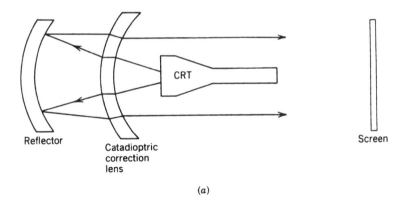

(a)

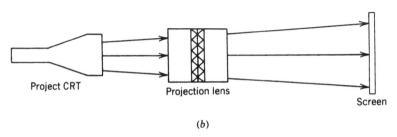

(b)

Figure 3.52 Direct projection system: (a) reflective projection CRT system; (b) transmission projection CRT system. After Gibilini and House [22], by permission Pennwell Publishing Co.

lens axis is always at an angle to the screen for most systems of this type. As a result, the plane of the phosphor surface must be tilted in order to maintain focus across the screen. This is known as the *Scheimpflug condition* and is illustrated in Figure 3.53. However, this requirement leads to TCL variation across the screen, and causes *keystoning*, or the conversion of a rectangle on the CRT surface into a trapezoid on the screen. Keystoning occurs when a rectangle on the phospher surface appears as a trapezoid on the screen. This effect is minimized by using a long TCL and a shorter angle between the lens and the screen.

3.5.3 Light Valve and LCD Projection

3.5.3.1 Basic System

Light-valve projection systems overcome some of the deficiencies of CRT projection systems, in particular limited screen size and output luminance. *Light valves* may be defined as any devices that control light intensity by varying some optical characteristics of the transmission media. One basic

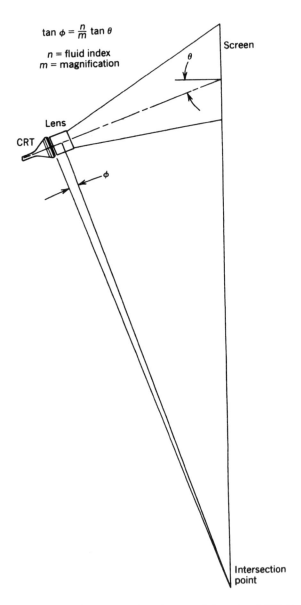

$\tan \phi = \frac{n}{m} \tan \theta$

n = fluid index
m = magnification

Screen

θ

Lens

CRT

ϕ

Intersection
point

Figure 3.53 Orientation of screen, phosphor, and lens to fulfill the Scheimpflug condition. After Moskovich et al. [23], by permission of SID.

approach to achieving such a device, illustrated in Figure 3.54, consists of a light source, special optics known as a *Schlieren system*, some type of deformable medium, a means for controlling the amount and type of deformation, and a final projection lens. The light source may be a high-intensity type that allows the final output to be of sufficient magnitude to achieve high levels of luminance at the screen. This light passes through the first set of

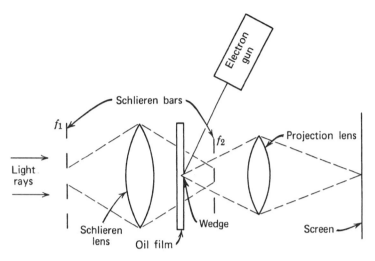

Figure 3.54   Refractive oil film light-valve system. After Sherr [24, p. 211]. Reprinted by permission of John Wiley & Sons, Inc.

Schlieren bars and is focused by the Schlieren lens on an oil film that is laid on a transparent plate. The electron gun is arranged so that the beam of emitted electrons strikes the film and causes the film to be distorted to an extent and in a pattern determined by the shape of the beam. Light in this shape then passes through the second set of Schlieren bars and is focused on the screen through the projection lens. The electron beam may be caused to scan the film in a raster format so that the resultant image is in the form of a TV image. When the oil film is smooth, no light passes through the bars, but when it is deformed by the scanning beam in the form of a TV raster, the ripples refract the light, causing it to pass through the second set of bars in the shape of a raster that then appears on the screen. The magnitude of the light is proportional to the depth of the ripples because the angle of refraction is proportional to the depth, and the depth is proportional to the energy in the beam.

   This is a relatively simple light-valve system, but it is difficult to obtain good contrast ratios with this design because sizes and spacing of the two Schlieren bars must be maintained to a very tight tolerance. In addition, the ratio between maximum and minimum outputs to the screen is small because only small defraction angles can be achieved. These difficulties are overcome by using a diffraction grating technique in which the electron beam creates a diffraction grating rather than a refraction deformation on the oil film. This approach, shown in Figure 3.55, has been incorporated in the commercial system known as the *Eidophor* (see Fig. 3.56), in which the electron beam that creates the diffraction grating is in the form of a magnetically deflected and focused electron gun, subject to the same equations 2.8–2.10 as the CRT beam. It should also be noted that only a single set of bars is used, with the

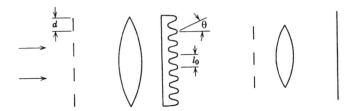

Figure 3.55 Diffraction grating oil film light-valve system. After Sherr [24, p. 211]. Reprinted by permission of John Wiley & Sons, Inc.

light beam passing through the bars twice. Resolution of over 1000 TV lines and luminance above 100 fL have been achieved with corresponding high-quality video images. Therefore, this design, in spite of certain problems with oil-film contamination and the special chemical characteristics required for this film, has been successfully designed and manufactured, and is in use in a variety of applications, in particular for instant replay in sports arenas.

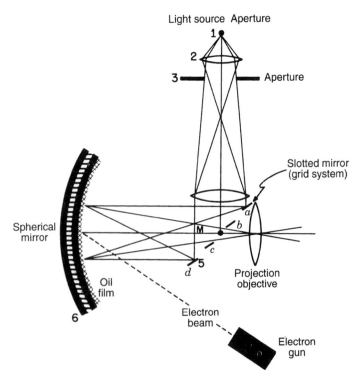

Figure 3.56 Eidophor light-valve projection. After Horowitz [25], by permission of SID.

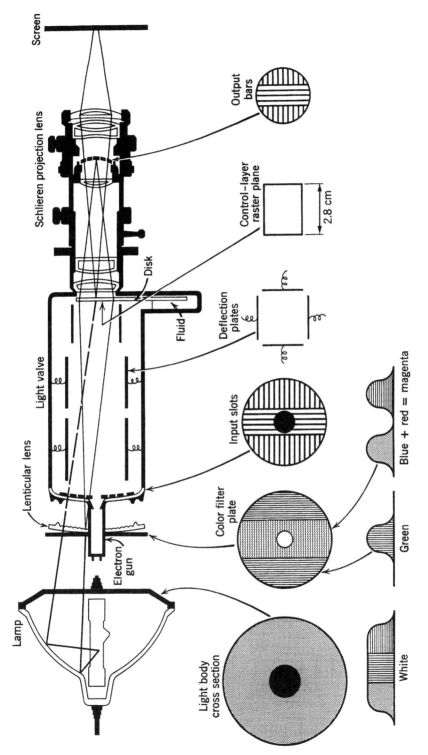

Figure 3.57 Sealed light-valve schematic. After Good [26], by permission of SID.

159

However, its cost is quite high, especially as three systems are required for color. This has led to several attempts to design systems that are somewhat less complex. One such attempt by the General Electric Company has resulted in the ingenious system shown in Figure 3.57, where the entire light valve is placed in a sealed envelope and the three gratings for the three colors are placed on a single oil film, thus obviating the three independent systems required by the color Eidophor. This simplification has resulted in a somewhat lower-cost color light-valve system, although this advantage was reduced when it was found necessary to add another sealed light valve to achieve the higher resolution required for non-NTSC TV displays.

3.5.3.2   *Liquid Crystal*

Although several light-valve projection systems use LC technology, the main types of projection systems based on LCDs are those that use LC panels as slides that, when combined with standard or special slide projectors and computers, result in projection systems that are dynamic in their information capabilities, much as are the CRT and light-valve types, but with many of the desirable characteristics and only a few of the deficiencies of both of the others, in particular, limited resolution and output luminance. The basic system configuration for a representative projection system is shown in Figure 3.58, including a dedicated slide projector. This system contains three

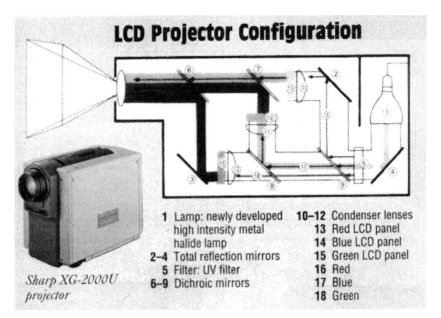

Figure 3.58 LCD projector configuration. Reproduced by permission of Sharp Electronics Corporation.

TABLE 3.32   LCD Projection Panel Parameters

Parameter	Active Matrix	Passive Matrix
Colors (no.)	1.7 million × 1–16 million	512
Resolution (pixels $W \times H$)	640 × 480	640 × 480
Contrast ratio	60: 1–130: 1	18: 1
Video input	NTSC/PAL/SECAM	No
Size (diagonal in.)	8.4–10.4	9.6
No. LCD panels	1–3	1

AMLCD panels, one for each color, and a high-intensity light source, configured much as for the light-valve systems. The dynamic LCD slide panels are essentially the same as the standard LCD panels, and are still subject to the limitations in performance found in the panels, as is evident from the parameter values given in Table 3.32. However, rapid improvements in performance are anticipated. These are representative parameters for projection panels, and other combinations are also possible and available, as are complete projectors with a 250-W halide lamp as a light source that allows an initial luminance of up to 800 lux to be produced, on screen sizes of up to 300 ft. This does not as yet meet light-valve system capabilities, but is competitive with CRT units.

3.6   SUMMARY AND CONCLUSIONS

This completes the review and discussion of the major display products that are used for the various applications listed in Tables 1.1 and 1.2. At this point it is useful to list those applications again with the products and technologies. This is done in Table 3.33, which may be used as a guide to the relationship between product types and applications.

It is apparent that there are various choices of both technologies and products for each application. Therefore, it is useful to attempt an evaluation rating for each application in terms of these choices. This is done on the basis of using a range of 1–5, with 5 the best and 1 the worst; the results are shown in Table 3.34. It should be noted that a rating of 1 does not indicate that either the technology or the product is useless, merely that it is not as good as other choices with higher ratings.

Of course, these evaluations apply only to operational capabilities for the specific application, and other factors such as cost, availability, portability, and similar considerations may be involved to determine the final choice. For example, although CRT is the highest-rated technology in every instance, for portable electronic games, FPDs, in particular those using the LCD technology, are to be preferred. However, to include all the other possible factors in

TABLE 3.33   Applications, Technologies, and Products

Application	Technology	Product Types
Advertising	CRT, IN, EM, LED, LCD	Monitor, large-board
Animation	CRT, EL, LCD, plasma	Monitor, FPD, projection
Art	CRT, LCD, inkjet, laser, pen, thermal	Monitor, FPD, plotter
Avionics	CRT, LCD	Monitor, FPD, projection
Communication	CRT, LED, LCD	Monitor, readout, FPD, large-board
CAD/CAM/CAE/CIM	CRT, EL, plasma, LCD, inkjet, laser, pen, thermal	Monitor, readout, FPD, projection
Computer graphics	CRT, EL, plasma, LCD, inkjet, laser, pen, thermal	Monitor, FPD, projection
Control and monitoring	CRT, LED, LCD	Monitor, readout, FPD, large-board
Desktop publishing	CRT, LCD, EL, plasma, pen, inkjet	Monitor, FPD, plotter
Education	CRT, LCD	Monitor, FPD, projection
Electronic games	CRT, EL, plasma, LCD	Monitor, FPD
Geographic systems	CRT, EL, plasma, LCD	Monitor, FPD, plotter, projection
Home entertainment	CRT, LCD	Monitor, FPD, projection
Imaging systems	CRT, EL, plasma, LCD	Monitor, FPD, projection
Information systems	CRT, EL plasma, LCD, inkjet, pen	Monitor, FPD plotter, projection
Instruments	CRT, LED, LCD, VFD	Monitor, FPD, readout
Manufacturing	CRT, LED, VFD, EL, plasma	Monitor, FPD, readout, large-board
Measurements	CRT, LCD, VFD	Monitor, readout, FPD
Military	CRT, LCD, EL, plasma	Monitor, FPD, projection
Multimedia	CRT, EL, LCD, plasma	Monitor, FPD, projection
Navigation	CRT, LCD, VFD	Monitor, FPD, readout
Presentations	CRT, LCD	Monitor, FPD, projection
Simulation	CRT, EL, plasma, LCD	Monitor, FPD, projection
Sports	Light-valve, incandescent	Projection, large-board
Television	CRT, LCD	Monitor, projection
Transportation	CRT, LCD, electromagnetic	Monitor, FPD, large-board
Utilities	CRT, LCD, electromagnetic	Monitor, FPD, readout, large-board
Virtual reality	CRT, LCD	Monitor, FPD, projection

TABLE 3.34   Applications versus Ratings

Application	Technology	Rating	Product	Rating
Advertising	CRT, LCD	5, 4	Monitor, FPD	5, 4
	EM, IN, LED	2, 2, 3	Large board	3
	LCD	4	FPD, monitor	4, 5
Animation	CRT	5	Monitor	5
	EL, LCD	4, 5	FPD, monitor	4, 5
	Plasma	4	FPD, monitor	4, 4
Art	CRT, LCD	5, 4	FPD, monitor	4, 5
	Inkjet, laser	5, 4	Plotter	5
	Pen, thermal	4, 5	Plotter	5
Avionics	CRT, LCD	5, 5	FPD, monitor	5, 5
			Projector	4
Communication	CRT, LED,	5, 5	Monitor, readout	5, 4
	LCD	4	FPD	4
CAD/CAM/CAE/CIM	CRT, EL	5, 3	Monitor, FPD	5, 3
	Plasma, LCD	3, 4	FPD, projector	3, 4
	Inkjet, laser	4, 3	Printer, plotter	4, 5
	Pen, thermal	3, 4		
Computer graphics	CRT, EL	5, 3	Monitor, FPD	5, 3
	Plasma, LCD	3, 4	FPD, projector	3, 5
	Inkjet, laser	4, 3	Plotter	5
	Pen, thermal	3, 4	Plotter	5
Control and monitoring	CRT, LED	5, 3	Monitor, readout	5, 3
	LCD	4	FPD, projector	4, 4
Desktop publishing	CRT, EL	4, 3	Monitor, FPD	5, 4
	LCD, plasma	4, 3	FPD	4
	Inkjet, pen	4, 3	Printer, plotter	4, 4
Education	CRT, LCD	5, 4	Monitor, FPD	5, 4
			Projector	4
Electronic games	CRT, EL	5, 3	Monitor, FPD	5, 3
	Plasma, LCD	3, 4	FPD	4
Geographic information	CRT, EL	5, 3	Monitor, FPD	5, 3
systems	Plasma, LCD	3, 4	FPD	4
	Inkjet, pen	4, 3	Plotter	4
Home entertainment	CRT, LCD	5, 3	Monitor, FPD	5, 3
Imaging systems	CRT, EL	5, 2	Monitor, FPD	5, 3
	Plasma, LCD	3, 3	Projector	4
Information systems	CRT, LCD	5, 2	Monitor, FPD	5, 3
	Inkjet, pen	4, 3	Plotter, projector	4, 4
Instruments	CRT, LED	5, 3	Monitor, readout	5, 5
	LCD, VFD	4, 4	Readout	5
Manufacturing	CRT, EL	5, 3	Monitor, FPD	5, 4
	LED, plasma	5, 3	Readout, FPD	5, 3
	VFD	5	Readout	5
Measurements	CRT, LCD	5, 4	Monitor, FPD	5, 4
	VFD	5	Readout	5

TABLE 3.34   (*Continued*)

Application	Technology	Rating	Product	Rating
Military	CRT, EL	5, 4	Monitor, FPD	5, 4
	LCD, plasma	5, 3	FPD, projector	4, 3
Multimedia	CRT, EL	5, 3	Monitor, FPD	5, 3
	LCD, plasma	4, 3	Monitor, FPD	4, 3
			Projector	5
Navigation	CRT, LCD	5, 4	Monitor, FPD	5, 4
	VFD	4	Readout	5
Presentations	CRT, LCD	5, 4	Monitor, FPD	5, 4
	CRT, LCD	5, 5	Projector	5
Simulation	CRT, EL	5, 3	Monitor, FPD,	5, 4
	LCD, plasma		FPD, projector	4, 5
Sports	Light–valve, IN	5, 3	Projector, board	5, 3
Television	CRT, LCD	5, 4	Monitor, projector	5, 4
Transportation	CRT, LCD	5, 4	Monitor, FPD	5, 3
	LED, IN, EM	3, 3, 4	Large board	4
Utilities	CRT, LCD	5, 3	Monitor, FPD	5, 3
	EM	4	Readout, board	4, 4
Virtual reality	CRT, LCD	5, 4	Monitor, FPD	5, 5
			Projector	5

the ratings would be quite unwieldy, and this text is concerned primarily with the technical performance characteristics, so the rating are restricted to those. The rest is left to the student, that is, the reader of this tome.

This brings us to the end of this chapter on output devices as products in the context of electronic display equipment and systems. It covers essentially all of the available and some of the ones expected to be available in the foreseeable future. This includes a large number of such products and some may have been overlooked or neglected in this treatment. If so, it was done without malice, and may be corrected in a future edition.

REFERENCES

1. Gage, S., et al., *Optoelectronics/Fiber-optics Application Manual*, 2nd ed., Mc-Graw-Hill, New York, 1981.
2. Cola, R., et al., "Gas Discharge Panels with Internal Line Sequencing (Self-Scan[R])," in B. Kazan, Ed., *Advances in Image Pickup and Display*, Academic Press, New York, 1977.
3. Birk, J. D., "Emissives Get Brighter-with Colors," *Inform. Display* **10**/12, 1994.
4. King, C. N., "Electroluminescent Display Systems," *SID Sem. Lect. Notes* 1994, M-9.

5. Khormael, R., et al., "High-Resolution Active-Matrix Electroluminescent Display," *SID 1994 Dig.* 138.

6. (a) Vecht, A., "AC and DC Electroluminescent Displays," *SID Sem. Lect. Notes* 1990, F-2. (b) King, C. N., op. cit.

7. Uchiike, H., "Color Plasma Displays," *SID Sem. Lect. Notes*, 1991 M8/11–M8/19.

8. (a) Weber, I. F., "Plasma Displays," *SID Sem. Lect. Notes*, 1994 M-8/31; (b) Lieberman, D., "New Twist in LCD Designs Makes Them Thin, Light Power Savers," *Comput. Design April* 1, 1989, 37.

9. Conner, A. R., and Gulick, P. E., "High Resolution Display System Based on Stacked Mutually Compensated STN-LCD Layers," *SID 1991 Dig.* 755.

10. Scheffer, T. J., and Nehring, J., "Supertwisted-Nematic (STN) LCDs," *SID Sem. Lect. Notes* 1994, M-1.

11. Xie, C., "Field-Emission Characteristic Required for Field-Emission Displays," *IDRC 1994 Conf. Rec.*, 444.

12. (a) Santelmann, W. F., Jr., and Mason, D. P., "Managing the Beam in CRT Displays," *Inform. Display* **7**/1, 1991, 17; (b) Sherr, S., *Video and Electronics Displays: A User's Guide*, Wiley, 1982; (c) King, C. N., op. cit.

13. Conner, A., "The Evolution of the Stacked Color LCD," *SID Appl. Notes* **92**, 110.

14. Schlam, F., "Overview of Flat Panel Displays," *SID Sem. Lect. Notes* 1991, F-1/30.

15. Patterson, M. L., and Lynch, G. L., "Development of a Large Drafting Plotter," *Hewlett-Packard J.* Nov. 1981, 3.

16. Mills, R. N., "Color Hard Copy," *SID Sem. Lect. Notes* 1989, 7.28–7.39.

17. Jaffe, A. B., and Burland, G. M., "Electrophotographic Printing," in R. C. Durbeck and S. Sherr, Eds., *Output Hard Copy Devices*, Academic Press, New York, 1988.

18. Sahni, O., "Color Printer Technologies," *SID Sem. Lect. Notes* 1990, F-7-9.

19. Williams, R. A., "Wire Matrix Printing," in R. C. Durbeck and S. Sherr, Eds., *Output Hard Copy Devices*, Academic Press, New York, 1988.

20. Wurtz, J., "CRTs for Hard Copy," in R. C. Durbeck and S. Sherr, Eds., *Output Hard Copy Devices*, Academic Press, New York, 1988.

21. Pender, H., and McIlwain, K., Eds., *Electrical Engineers' Handbook*, Vol. 2, *Communication-Electronics*, 4th ed., Wiley, New York, 1950.

22. Gibilini, D., and House, W. R., "CRTs Project High Definition Television Pictures," *Laser Focus World* Sept. 1990, 125.

23. Moskovich, J., et al., "CRT Projection Optics," *SID Sem. Lect. Notes* 1991, M7/17–M7/18.

24. Sherr, S., *Fundamentals of Displays System Design*, Wiley, New York, 1970.

25. Horowitz, P., "Concept for Design and Implementation of Mobile Computer-Generated Display System," *Inform. Display* **3**/4, 1966, 27.

26. Good, W. F., "Projection Television," *SID Proc.* **17**/1, 1976, 3'7.

4

SYSTEMS AND USER APPLICATIONS

4.1 INTRODUCTION

Systems discussed here include those data processing units, terminals, and workstations equipped with display devices or systems as the basic means for presenting output data. Foremost among these are the many different categories of computers, ranging from the most elaborate mainframes to the simplest palmheld types, and even some of the more elaborate game playing systems. Indeed, when equipped with the appropriate output means and operating programs, the desktop personal computer can qualify as both a terminal and a workstation, the difference being that the output means are intrinsic parts of the latter, whereas they must be added to the basic computer. However, although in many cases this is more a distinction than a real difference, the designations of personal computer, terminal, and workstation are still in use, and are treated separately here in terms of the display aspects of the units. This leads to some repetition but is useful in describing the attributes of the three major types of data processing systems. These systems are treated consecutively, beginning with computers, followed by the various types of terminals, and ending with workstations.

Following these discussions, the relative merits of the different system types, as they apply to the various applications briefly described in Chapter 1, are reviewed and evaluated, with performance specifications provided for each major type. This is followed by an examination of the relative merits of workstations and host terminal systems as they apply to the requirements of a group of user applications introduced and briefly described in Chapter 1. These earlier descriptions are expanded and supplemented by operating electronic display specifications for the applications covered in this chapter.

Before embarking on separate descriptions of these equipments, systems, and applications, it is useful to present some basic block diagrams for what may be considered as a display subsystem in the context used in this volume.

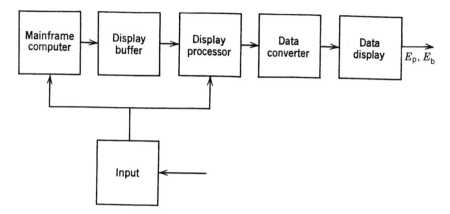

Figure 4.1 Generalized display system block diagram. After Sherr [1], by permission Wiley.

These are shown in Figure 4.1 which presents a generalized display system, and Figure 4.2, which includes several more complex examples of such systems that add different amounts of processing to the basic system. A full workstation is shown in Figure 4.3, where the graphics subsystem may also be defined as an intelligent terminal that may be used in either a standalone configuration, or connected to a host computer. Thus, the host computer, usually a mainframe, or a mini used as a server, can have a workstation as its display subsystem or some simpler terminal. In any event, these systems and

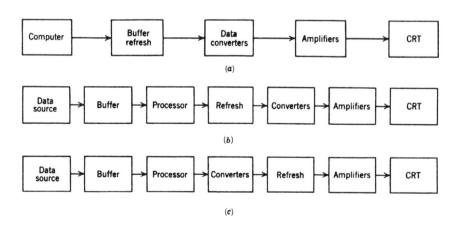

Figure 4.2 Expanded workstation block diagram. (*a*) Minimum processing; (*b*) extensive processing; (*c*) digital television. After Sherr [2], by permission Wiley.

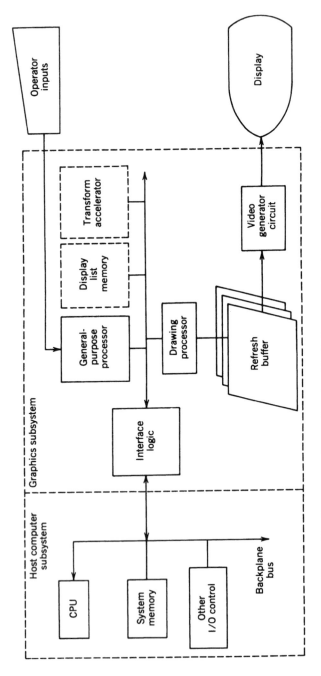

Figure 4.3 Representative workstation block diagram. After Breen [3], by permission SID.

subsystems range from the simplest ASCII (American Standard Code for Information Interchange) (dumb) terminal, which is little more than a monitor with minimal electronics through the more complex workstations that may be made up of a personal computer and peripherals, through an elaborate system consisting of a host computer and an advanced-capability workstation. Each of these systems and equipment components is considered in detail next, followed by the section on applications.

4.2 COMPUTERS

4.2.1 Mainframes and Minis

Mainframe computers take a much smaller share of the market than was the case in the not-too-distant past. Indeed, most of their functions have been taken over by client servers, in particular the RISC (reduced instruction set computer)-based versions, and even by some of the more powerful desktops. It may be anticipated that mainframes will either ultimately disappear in their present form, or be limited to supercomputers, although even the latter are being replaced by the massively parallel units. However, there are a number of mainframes still in use, with others being obtained, and it is of some interest to examine what types of displays are available and provided for these data processing systems. These displays consists of hard-copy devices, as the outputs are generally put onto some type of semipermanent form, and dynamic displays such as CRT monitors and terminals. The hard-copy devices are usually page printers, although in some cases the data are presented as a plot and a plotter is used. In addition, some very elaborate printing systems are available, as well as a variety of special-function printers, such as one that prints magnetic ink character recognition. Thus essentially any generally available hard-copy type of display may be obtained to work with the basic mainframe. In this respect, mainframe computer systems provide the broadest range of output hard-copy units and are more appropriate for those applications that call for a large amount of hard copy such as business systems that handle payroll and other personnel-oriented functions. Mainframes have also played a role in the various computer-aided design and engineering applications, as well as the computer-aided and integrated manufacturing, and hard copy plays an important role at least in CAD and CAE, and a lesser role in CAM. Some of these hard-copy units are listed in Table 4.1.

Insofar as the CRT units are concerned, the monitors are essentially the same as those discussed previously (Section 3.3.2) but the special-purpose terminals are of more significance. These will generally include an interface to the mainframe and a small amount of data processing to add to the ability to view what is going on in the mainframe. For example, a logic unit and keyboard in conjunction with a monitor allows both the PC and mainframe

TABLE 4.1   Hard-Copy Units for Mainframe Computers

Unit	Type	Description
Page printer	Electrophotographic	Laser: 12–58 (ipm) (impressions per minute)
		LED: 30 ipm
Function printer	Electrophotographic	Laser: 91 ipm
		LED: 92 ipm
Printer	Laser	10 ppm
Impact printer	Engraved band	1200–2200 ipm
	Dot-matrix	200–600 cps (cycles per second)
Plotter	Pen	0.001 resolution
Plotter–printer	Inkjet	360 × 720 resolution
	Electrostatic	200 dpi (dots per in.)

host to be labeled on the keys, and other units may provide a printer port and on screen calculator in addition to a view of the host system.

Looking further at the mini group, it is being at least partially replaced by the high-performance PCs and workstations that can accomplish most of what minis can do. However, some versions, such as RISC types that seem most compatible with the mini architecture, are retaining their place in the market. The display portions of these systems add nothing to those found in mainframe systems, and need no further discussion at this point. These systems use some of the terminals and workstations covered in the following sections, and discussions are deferred to those sections.

4.2.2   Desktop, Laptop, Notebook, and Palmtop Computers

These computers constitute the most rapidly growing segment of the computer industry, and threaten to take over almost all of the functions previously carried out by mainframes and minis. They also use the greatest variety of displays, including a wide array of monitors, FPDs, and hard-copy devices. The desktop types have shown the highest performance capabilities, but are being overtaken by the portable types, in particular laptops and notebooks, although these are still behind the former in total capability. The use of CRT monitors for desktops exclusively has helped to maintain performance capabilities related to the higher resolution still found in monitors that use CRT technology. However, some of the laptop types include the ability to drive external monitors, which provides them with equivalent performance, albeit for increased cost. Some of this range is shown in Table 4.2, where the computer types are matched with what appears to be the most appropriate displays for each specific type are present. This choice, of course, is subject to change based on improvements in available types and other factors such as price and availability.

It should be noted that the desktop type with a CRT monitor offers the best performance in terms of display size and resolution, but the portables are beginning to approach this performance, with 15-in. color panels in

TABLE 4.2   **Personal Computer / Workstation versus Display Type**

Parameters	Desktop	Laptop	Notebook	Subnotebook	Palmtop
Display type	Monitor	Monitor, FPD	FPD	FPD	FPD
Technology	CRT	EL, LCD	LCD	LCD	LCD
Display size (in.)	12–17	7.7–10.3	7.7–10.3	8.4–9.5	3–5
Resolution	640 × 480–	640 × 480–	640 × 480–	640 × 480	640 × 480
	1280 × 1024	1024 × 768	1024 × 768	1024 × 768	
Weight (lb)	NA	4–8	3.5–6	3.5–6	< 3
Battery life (h)	NA	4–8	2.5–8	3–6	2–4

advanced development. However, CRT color is still the best and may be a deciding factor when determining which computer is best for the application. Of course, when portability is a must, then the battery-operated types remain the only choice with the accompanying disadvantage of low battery life. This is partially ameliorated by the ease of battery replacement and recharging, but may be a problem when the application requires the user to be remote from a charging facility, so spare batteries must be included. Some of the visual limitations of the FPDs may be overcome by using a CRT monitor, but this limits the use to applications where ac power is available, and the only gain is the smaller size and power requirements of the portable units. However, changes in FPDs and portable computers are occurring at a rapid pace and may lead to more attractive choices in these units in the near future.

4.3 TERMINALS AND WORKSTATIONS

4.3.1 Terminals

Various terminals operate with data processing capabilities ranging from minimal to moderate, and use some type of computer as the source of the required additional capabilities. Most prominent among these types are those termed (appropriately) X or X Windows terminals, and those designated as ASCII or ANSI (American National Standards Institute) terminals. A typical example of an X Windows terminal architecture, shown in Figure 4.4, contains some of the capabilities found in workstations but is essentially a diskless workstation that lacks some of the hardware found in the latter such as large memory management units. Therefore, X Windows terminals are cheaper than workstations, and may be considered as a workstation designed for a specific application, which operates under an X Windows protocol. The place of such terminals in a typical host windows manager environment is shown in Figure 4.5, and parameters for an X Windows terminal in Table 4.3.

It is clear that a very high-quality CRT monitor is the basic output unit, to which might be added various of the printers and plotters that come with many of the computer systems described previously. In addition, as shown in Figure 4.5, both a host mainframe and a client–server may be part of the

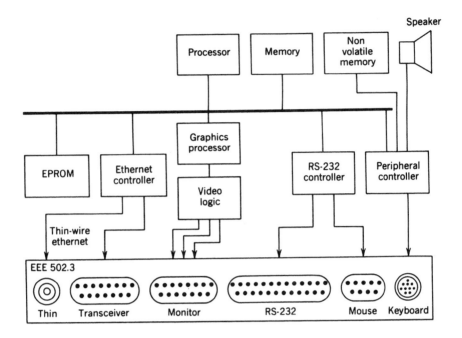

Figure 4.4   Typical first-generation X terminal architecture. After Wilson [4], by permission of *Computer Design* (Aug. 1991, p. 78).

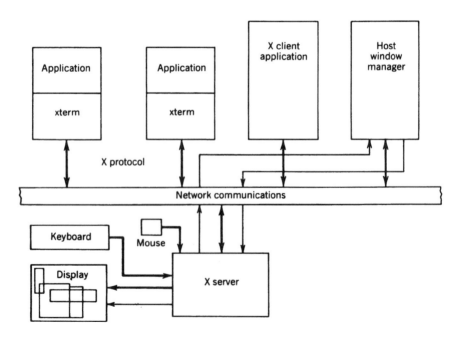

Figure 4.5   Typical host window manager environment. Courtesy of NCR Corporation.

TABLE 4.3   X Windows Terminal Specifications

Parameter	Value
Processor speed (MHz)	16–22.7
User memory (Mbytes)	4–18
Video memory (Mbytes)	2
Monitor type (diagonal in.)	14–21 color
Resolution	800 × 600–1280 × 1024
Colors (No.)	256
Color palette	16.7 million
Refresh rate (Hz)	60–78

total operating system. These data processing units may also call for other types of dynamic output units, but the terminal is usually adequately served by the CRT monitor type with the specifications listed in Table 4.3.

Finally, among the available types of terminals is the simplest one, referred to previously as the ASCII unit, and called "dumb," but now dignified by the somewhat more prestigious designation used here. These are terminals with a minimum of data processing capabilities that require connection to some other data processing source to make them useful. Their limited amount of standalone capability is demonstrated by the specifications given in Table 4.4 and illustrated in the block diagram shown in Figure 4.6. This is the type of terminal with the absolute minimum capability to be found in any large number of installations and is relatively inexpensive as an output unit. Of course, the host computer or computers are more expensive than even the higher-priced desktop units, but the combination of computer and low-cost terminals can result in a lower total cost for equivalent total capability. The recent proliferation of client–server applications has made this terminal type with its associated computer and software one of the most rapidly growing application groups that use data processing systems as the source of the operational activity. This growth makes it of interest to examine these applications in some detail, particularly in terms of what the required display parameter values need to be, and how well they can be met by the different types of monitors or display panels that might be used. However,

TABLE 4.4   ASCII Terminal Specifications

Parameter	Value
Phosphor colors	G, A, W
CRT (diagonal in.)	14
Compatibility	ASCII, ANSI
Refresh rate (Hz)	60–78
Character resolution	10 × 12–16 × 208
Columns	80–161
Rows	26–44
Process chip type 68,000	8 MHz

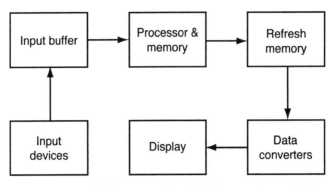

Figure 4.6   Block diagram of dumb alphanumeric terminal.

prior to embarking on such a discussion, it is advisable to complete this review of the characteristics of the types of terminals used for the client–server system, as well as other application groups. This may be done by presenting a representative set of parameters for ASCII/ANSI terminals, as shown in Table 4.4, and in conjunction with those for X terminals given in Table 4.3, essentially defining the basic display parameters. Then, some combination of host server and one or more of the terminal types may be used to provide the capability required by the client user of this client–server system. Of course, software plays a significant role in determining where the data processing operations are performed, and therefore what type of host computer and terminal type make up the best combination. These units have much in common with the X terminals, except for processor speed, memory, and data processing capabilities. They are generally limited to A/N data, but it is also possible to obtain units with some graphics capabilities. In any event, between these units and the X terminals there is ample room for a wide variety of applications when some type of host computer or server is available. Client–server applications are those that are met by a combination of a separate data processing units and the types of terminals discussed previously. In certain respects, the resultant system is somewhat similar to the earlier general-purpose host–terminal combinations that was so success-fully served by the mainframe–terminal combinations in which IBM played a leading role. The applications and terminal types are listed in Table 4.5. These are just a few of the many possible applications that use one or more of these terminal types as the output system driven by some form of host computer. The configuration shown in Figure 4.5 is representative, with client–server applications handled by the host. One advantage of these combinations is that the host computer can off-load certain data processing functions, and the total system may be less costly than an equivalent group of PC units that can provide the same total functional capabilities. In any event, there is a market for these systems, including the terminals, which should continue to exist for some time.

TABLE 4.5   Client–Server Applications

Application	Terminal Type
Process control	X Windows
Desktop publishing	X Windows
CAD/CAM	X Windows
Database management	ASCII
Word processing	ASCII
Document preparation	ASCII, X Windows

4.3.2   Workstations

Workstations are the most powerful and elaborate of the different types of operating systems that are used in conjunction with, and in many cases independent of, host computers. A representative workstation block diagram is shown in Figure 4.7, and it is apparent from this illustration that workstations can be operated in either standalone or host-driven configuration. Indeed, the high-level workstation type illustrated by the block diagram in Figure 4.3 can and does operate largely in a standalone mode, with the capability of interfacing with a variety of sources. This type of unit is particularly adapted to graphics operations and can be used in conjunction with a variety of monitors. The graphics capabilities are aided by the special elements shown in Figure 4.3 such as the special processors and memories that are devoted to the display-related functions. Of course, some of these functions can be emulated by software contained in general-purpose micro-computers so that the difference between a standalone workstation and one that uses a host computer is blurred, and the same may be said about the workstations that contain high-performance monitors and those that can connect to separate monitors with high-performance capabilities. These choices are further demonstrated by the data given in Table 4.6. This table certainly demonstrates the wide performance range available in standalone workstations and makes them fully competitive with the host-based units. The choice is up to the user and is based primarily on cost considerations. However, in certain cases where cost is not the overriding factor, a valid argument may be made for selecting a high-level, standalone workstation because of its independence and ability to undertake unforeseen applications. It is also of interest to examine which applications are more appropriately performed at a workstation, when compared with the systems combinations that use host computers such as mainframes, minis, and client–server systems. To this end it is necessary to examine the applications in greater detail than has been done to this point, compare the capabilities required, and determine whether the extent to which they are made available by the host–terminal combination is better than the standalone workstation.

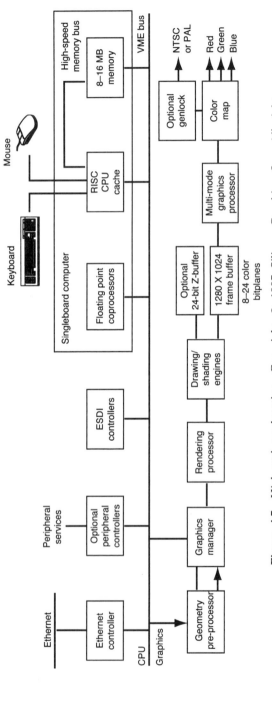

Figure 4.7 High-end workstation. Copyright © 1997 Silicon Graphics, Inc. All rights reserved.

TABLE 4.6   Workstation Specification

Parameter	Value
Processor speed (MHz)	25–100
Processor type	SPARC, MIPS, HP, RS
DRAM memory (Mbytes)	8–256
Performance (mips[a])	15–85
Internal display resolution	640 × 480–1280 × 1024
Internal display technology	CRT, LCD
Hard drive (Gbyte)	0–1

[a]Million instructions per second.

4.3.3   Input Devices

4.3.3.1   Introduction

Input devices are an integral part of every workstation, as are the hard-copy output devices described in Section 3.4. Therefore, it is appropriate at this point to include some discussion of the technologies used for and characteristics of the most popular input devices. These are, in the approximate order of their introduction into computer systems and workstations, the A/N keyboard, the light pen, the trackball, the data or graphics panel, and the ubiquitous mouse. Of these, the last two have been most popular, especially for graphics-oriented workstations, but the light pen and trackball are coming back into use. Similarly—and, like the paper that was expected to disappear with the advent of computer systems, but remains at least the same if not more present—the A/N keyboard is always at least made available, if not in as extensive use as the other devices. However, its significance to systems that emphasize the graphics aspect is relatively small, especially with the advent of Windows 95 and the possibility with the return of Steve Job to Apple, of the continued success of the Macintosh, so emphasis is placed on the other devices, in particular the graphics panel and mouse. These are covered in the following sections in some detail, with lesser attention paid to the light pen and trackball, and essentially none to the keyboard.

4.3.3.2   Light Pen

The light pen is probably the earliest of the graphic input devices, dating back to the early 1960s. It was the first means for accomplishing manual input to a random-deflection CRT system, and fell out of favor when the switch to raster systems began as it was more difficult to use in such a system. However, most of the initial difficulties have been overcome, and it is back in limited favor, at least enough to warrant some description. "Light pen" is a misnomer for this device, as it has only the limited physical appearance of a pen, as can be seen from the photograph shown in Figure 4.8, and does not emit any light. However, it is light-sensitive, as indicated by the schematic

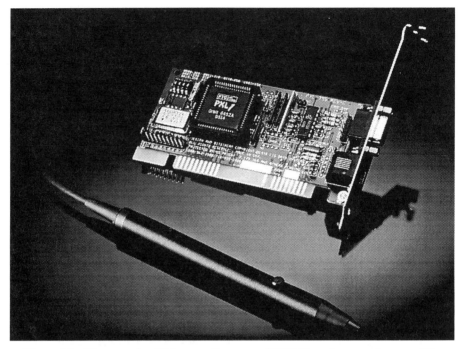

Figure 4.8 Photograph of light pen. Courtesy of FTG Data Systems, Stanton, CA.

shown in Figure 4.9. The phototransistor responds to the light input through the fiberoptics element, and produces a pulse of light at the output.

4.3.3.3 Graphics (Data) Panel

The graphics or data panel goes back in time to its initial conception in the mid-1960s as the Rand tablet, developed, as the name implies, at the Rand Corporation. A block diagram is shown in Figure 4.10, which illustrates the basic approach used for this as well as later data panel and digitizer designs. In this approach, a wand was used to pick up pulses from the X and Y pulse generators, and by counting the number of pulses in a given time period, one could determine the position of the wand on the panel surface. Subsequent designs use several other technologies, with the electromagnetic and sonic pulses the most popular. These are all digital designs, and the electromagnetic versions have become quite popular as digitizers as well as input devices.

4.3.3.4 Mouse

The mouse has become the most ubiquitous and popular of the input devices used in computer systems, and has advanced a considerable amount from the original design. It should be noted, in advance of further discussion, that the trackball preceded the mouse, and the latter might be termed an upside-down

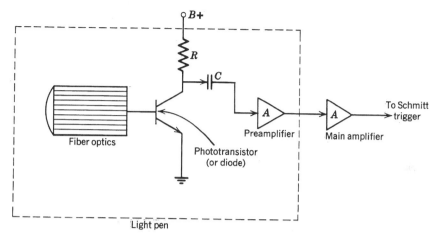

Figure 4.9 Light-pen schematic. After Sherr [5], by permission of John Wiley & Sons, Inc.

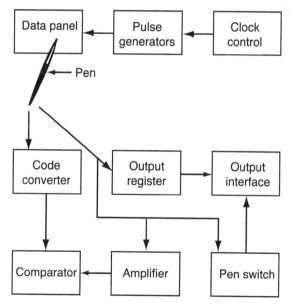

Figure 4.10 Data tablet block diagram.

trackball. However, the mouse does have some unique features of its own, and warrants separate attention because of its features and because so many users consider it the predecessor of the trackball. In any event, the basic operating structure of the original mouse design is pictured in Figure 4.11*a*, which demonstrates how the motion-sensing element, namely, the ball, causes the velocity of the ball circumference to equal the velocity of the mouse. This type of mouse is available in many shapes, as shown in Figure 4.11*b*, but each contains the motion sensor inside the shell and placed on a surface for operation. Then, as the mouse is moved on the surface, the ball rotates and transfers its motion to the shaft, where the motion is converted to either pulses or other electrical signals that can in turn be sent to the computer and used to move the cursor on the monitor screen. In addition to the all mechanical rotating system, there are optomechanical units in which the rotation is transferred by optical means, and cordless types that use infrared (IR) technology to transfer the motion data. In any event, regardless of the exact structural details, all of these units rotate a ball by moving the assembly with the ball on a flat surface, and convert the motion of the ball into electrical signals that are input into a computer and act to move a cursor at a rate and distance corresponding to the motion of the ball. There are also several buttons on top of the mouse, which are used to control the switches that determine whether the signals are transmitted. The three-button type has become more or less the preferred standard, especially now that Windows 95 uses the right button for a number of functions, as well as the left button for the usual mouse positioning actions. Insofar as the operating characteristics are concerned, the most important is resolution in terms of dots per inch (dpi), and this ranges from 200 to 400 and determines the resolution to which the cursor on the display screen can be placed, subject to the limitations imposed by the screen resolution.

4.3.3.5 Trackball

As noted previously, the trackball is the precurser of the mouse, although this fact does not appear to be generally known. Thus, it is amusing to note a trackball that is termed the "roller mouse," where the device is touted as an "upside-down idea [that] made the mouse obsolete." This device is pictured in Figure 4.12, and the close resemblance to an older version of the trackball shown in Figure 4.12*b* is quite apparent. Be that as it may, a number of new trackballs have become available, and it is in the throes of a rebirth. The trackball may use the same technology as the mouse; the major difference is

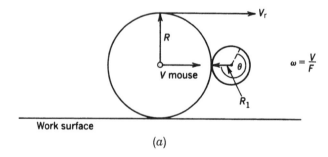

(*a*)

Figure 4.11 (*a*) Schematic of ball-and-shaft mouse structure. After Goy [6], by permission of Academic Press.

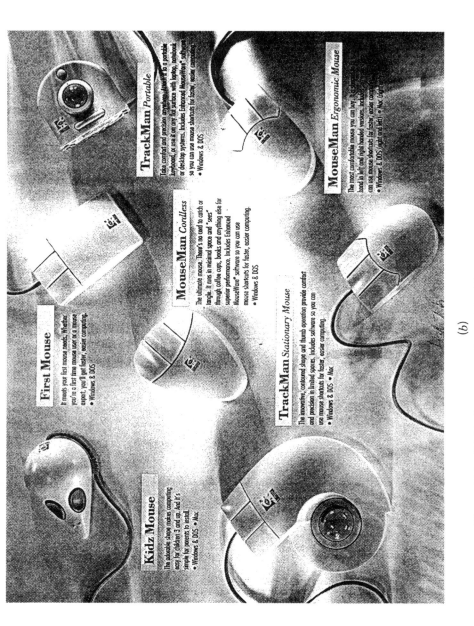

(b)

Figure 4.11 (b) Various mouse shapes. Used by permission. Logitech © 1997.

(*a*)

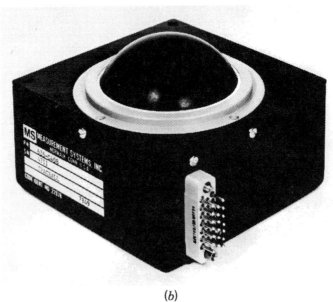

(*b*)

Figure 4.12   (*a*) Trackball Pro photograph. Courtesy of CH Products. (*b*) Photograph of standard 3-in. trackball. Courtesy of Measurement Systems, Inc.

that the ball is mounted on top and is available to the user to rotate while the assembly remains stationary, thus requiring much less desk space than the mouse. The output may consist of a series of pulses as for the mouse, and the cursor motion is the same. Thus, the two devices are essentially equivalent in their operation, and the user may make the choice depending on space, convenience, aesthetics, cost, and other factors that come to mind.

4.3.3.6 Other Input Devices

Numerous other input devices have achieved acceptance, albeit some of them quite fleetingly, and Table 4.7 lists most of these, along with the technology employed. Further details on these and the other four devices may be found elsewhere [6], but the previous discussions of the four main devices should suffice to introduce the user to the main attributes and characteristics of input devices used with the output display devices and systems.

4.4 WORKSTATIONS VERSUS HOST–TERMINAL SYSTEMS

4.4.1 Introduction

This section is devoted to the comparison of standalone workstations and host–terminal systems in terms of their operating capabilities as represented by parameter values, and the extent to which these capabilities satisfy the requirements of those applications for which both types are appropriate. As a first step, a list of relevant applications is given in Table 4.8 with the required operating parameters, particularly those that relate to the display aspects of the application requirements. This table is intended to supply the information necessary to arrive at some initial choice of operating parameters for each application listed. It is not intended to serve as a complete guide to preparing a specification or making a choice, but it does supply enough information about the required parameter values to allow at least a first, if somewhat tentative, choice to be made, which then can be followed up by a full specification. This is a fairly complete list of the major applications for data processing systems that include some type of electronic display, although it does not include all those applications, such as multimedia and virtual

TABLE 4.7   Input Devices

Device	Technology
Joystick	Straingauge (extensometer), resistive
Keyboard	Electromechanical, optomechanical
Touchscreen	Capacitive, resistive, light, piezoelectric
Pen	Electromagnetic, resistive
SAW	Surface acoustic wave (SAW)

TABLE 4.8   Application and Requirements

| Application | Processor Requirements | | Display Requirements | | |
	Speed (MHz)	Memory (Mbytes)	Resolution	Size (in.)	Color (no.)
Animation	66–100	120–480	1280 × 1024	21–27	256–16 million
Avionics	66–100	100–480	1280 × 1024	15–27	16–4096
Business graphics	20–33	40–120	1024 × 768	15–27	16–4096
Desktop publishing	20–33	40–120	800 × 600	15–21	256–4096
Communication	20–33	40–120	620 × 480	14–20	4–16
Information systems	8–16	120–480	800 × 600	14–20	16–256
Military	66–100	120–540	1280 × 1024	15–27	16–4096
Process control	15–50	15–50	620 × 480	14–20	16–256
Simulation	66–100	120–540	1280 × 1024	21–226	256–16 million

reality, for which the choice of the best data processing system is covered in the appropriate chapters. However, the discussion at this point establishes the basic procedure that is followed to arrive at the choice for essentially all possible applications and is followed in subsequent chapters devoted to user applications. One important application group is military and it may be assumed, as in Chapter 1, that the equivalent nonmilitary application also applies insofar as the choice of data processing systems is concerned, and the major difference is in the environmental and life characteristics rather than the operating characteristics covered here. With these provisos, we proceed to descriptions of the applications listed in Table 4.8 and an analysis of how they determine what characteristics are required of the systems.

4.4.2   Application Descriptions

4.4.2.1   Avionics

This is something of a catchall application in that it includes all the displays that might be used for any avionic requirement. The most significant example is navigation, which could possibly be included in the general category, but it is listed separately because there are some specialized types of avionics navigational displays that warrant some individual attention. The normal aircraft control panel has been a maze of instruments using a variety of numeric and A/N modules, as is shown in Figure 4.13, which are being combined into CRT displays that present a number of such displays on demand, as shown in Figure 4.14. Therefore, the tendency is to reduce the instrument panel into a minimum number of CRT displays that can be made to show the same information, not necessarily all at once. This is the trend, with the instrument panel a combination of several CRT or FPD units, with their specifications tailored to meet the requirements previously imposed on the individual instruments. The instrument-type displays are covered in the section on measurements below.

Engine instruments

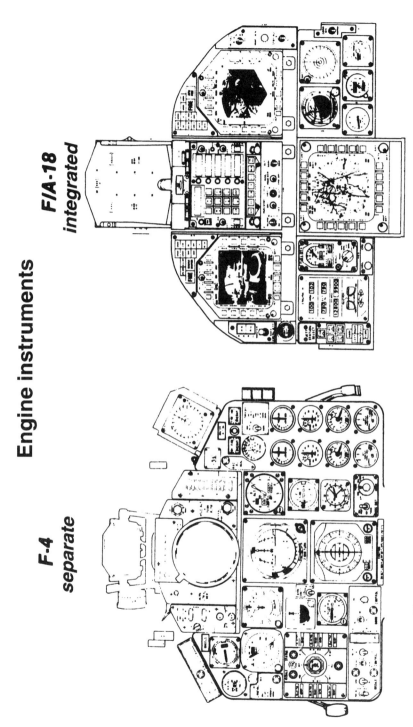

F/A-18
integrated

F-4
separate

Figure 4.13 Non-CRT aircraft control panel. After Adam [7], by permission of McDonnell Douglas Corporation.

Figure 4.14   CRT displays for an aircraft control panel. By permission of Thomson CSF.

In addition to these displays, there are several other navigation displays of greater complexity and correspondingly more demanding display specifications. These are the head-up (HUD), head-down (HDD), and helmet-mounted (HMD) displays. The HUD is a rather specialized type of display that presents a real-world display directly on the windshield, with a projected display from the HUD CRT that overlays the real-world image on the windshield. The form of the overlay image is shown in simplified form in Figure 4.15, and a functional diagram of the total system is shown in Figure 4.16. Holographic HUD displays have been produced in laboratory systems, but actual implementation is still in the future, and those displays are not considered further here. The HUD has also been proposed for automotive displays.

The HDD consists of two types: *horizontal situation display* (HSS) and *vertical situation display* (VSD). The HSS is used to display information about the orientation of the aircraft and other pertinent data about the orientation, as well as a variety of other data such as charts, electronically generated maps, and radar video, to name only a few types. The VSD is also a multifunction display that provides aircraft attitude information of the type shown in Figure 4.17 as well as information similar to that provided by the VSD but in relation to a plane perpendicular to the earth. Color and/or gray levels are desirable and should be part of the specification.

Finally, among these three navigation displays there is the HMD, the most sophisticated of these types. It consists of a helmet, worn on the head, which contains a display system that combines an angle and position sensing system

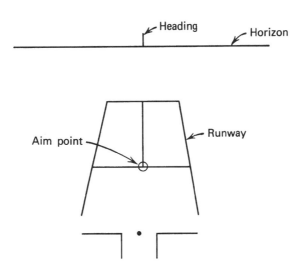

Figure 4.15 Characteristic data on head-up display. After Sherr [8], by permission of John Wiley & Sons, Inc.

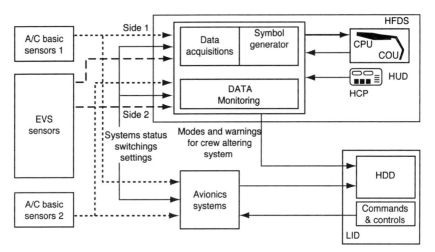

Figure 4.16 Head-up flight display system. After Perbet and Parus [9], by permission of SID.

with a high-resolution display and the optics needed to generate the image on a surface that can be viewed by the one who wears the helmet. The helmet therefore is a projection system where the projected information corresponds to what the wearer would see in a HUD with the addition of other images such as might be found on the HDD but in a form more immediately responsive to the viewing direction. Thus, it is a very powerful display system for both navigation and control since many of the operation controls can be slaved to the head movement. A cross section of one type of HMD is shown in Figure 4.18. The specifications for these three types are shown in Table 4.9, parts *a*, *b*, and *c*, respectively.

4.4.2.2 Business Graphics

Business graphics is broad classification that covers a range of requirements, depending on which of the many categories are to be covered. A fairly large number fall into this general category, and not all of them require the same amount of data processing capability or display performance. Thus, the system choice is dependent on how many of the possible applications may be supported by a specific system. In addition, it may be advisable to use several different systems with varied capabilities so that the most efficient combination may be achieved. This requires examining the needs of each application with some care so that they may be properly grouped and assigned to the most appropriate system. The major applications in the business graphics group are listed in Table 4.10, rated 1–10 in order of importance, with 10 the highest.

Maps

Electronic   **Film**

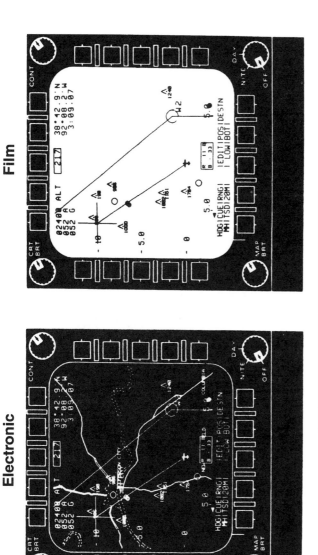

	Electronic	Film
Accuracy	OK	OK
Resolution	OK	Very good
Area coverage	1	X 100
Database	1995	Now

Figure 4.17 Electronic map display on aircraft panel. After Adam [10], by permission of McDonnell Douglas Corporation.

189

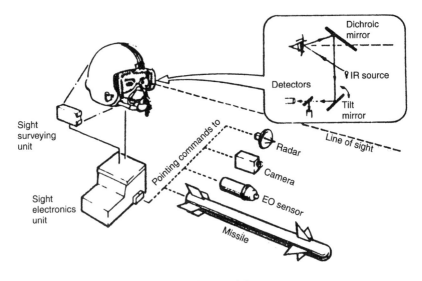

Helmet sight with eye pointing

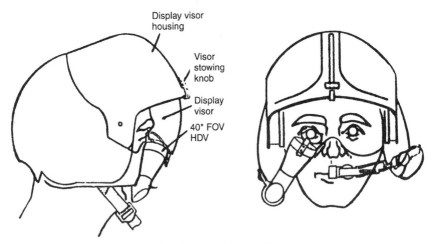

Honeywell helmet display unit

Figure 4.18 Helmet sight-and-display unit. Courtesy of McDonnell Douglas Corporation.

The system requirements that apply to all of these business graphics applications are listed in Table 4.11.

4.4.2.3 Communication
Communication is a fairly simple application as far as the end-user display system is concerned, consisting primarily of the telephone A/N readouts.

TABLE 4.9 Display Specification: Avionics Application

Parameter	Value	Technology	
		Input	Output
a. HUD			
Luminance (cd/m^2)	3500–6000	CRT, FPD, xenon	Screen
Contrast ratio (no.)	5–15	NA	NA
Resolution (TV lines)	525–1024	CRT, FPD	Screen
Speed (in./s)	1–10	CRT, FPD	Screen
Colors (no.)	1–4	CRT, FPD	NA
Viewing area ($H \times W$, in.)	3×4–6×9	CRT, FPD	Screen
Matrix (columns $\times$ rows)	400^2–800^2	FPD	Screen
Matrix element (in.)	0.002–0.01	FPD	Screen
b. HDD			
Luminance (cd/m^2)	20–45	NA	CRT, FPD
Contrast ratio (no.)	5–20	NA	NA
Resolution (TV lines)	525–1024	CRT	CRT, FPD
Speed (in./s)	1–10	CRT, FPD	CRT, FPD
Colors (no.)	4–256	NA	CRT, FPD
Viewing area ($H \times W$, in.)	6×10–10×15	CRT, FPD	CRT, FPD
Matrix (columns $\times$ rows)	$512^2 \times 1024^2$	NA	FPD
c. HMD			
Luminance (cd/m^2)	35–60	NA	CRT
Contrast ratio (no.)	5–15	NA	NA
Resolution (TV lines)	480–1280	NA	CRT
Speed (in./s)	1–10	CRT	CRT
Colors (no.)	16–256	CRT	CRT
Viewing area ($H \times W$)	80×60	CRT	CRT

TABLE 4.10 Business Graphics Applications

Application	Rating
Database management	10
Demographic data	8
Management presentations	10
Market analysis	7
Organization charts	5
Product investment planning	6
Product sales comparison	7
Project planning and control	10
Sales forecasts	9
Trend analysis	7

TABLE 4.11   Business Graphics System Requirements

Rating	Processor Requirements		Display Requirements		
	Speed (MHz)	Memory (Mbytes)	Resolution	Size (in.)	Color (no.)
10	33	120	800 × 600	26	4096
8–9	25–30	60–80	800 × 600	17–21	256–4096
5–7	20–25	40–60	640 × 480	15–17	16–256
10	—	—	High-end workstation		
8–9	—	—	Server/X terminal, workstation		
5–7	—	—	Server/ASCII terminal		

However, the central office may have a more elaborate display system using both terminals and large-board units. In addition, the data processing unit could be quite complex, with the ability to handle a large number of lines simultaneously. This could lead to a system with a client–server unit driving a multiplicity of ASCII terminals, with the addition of a few X Windows terminals connected to the same server, as well as several standalone workstations to cover those possible situations where the other terminals are not adequate. This is intended primarily for special failure conditions that require quick and accurate response and the means for correcting the condition by dispatching technical help to the affected area. The result of this combination of capabilities could lead to a rather elaborate and expensive system. However, in the majority of cases a much simpler system should be adequate with the more elaborate systems limited to only a few of the many central offices. In any event, the end-user systems are much simpler, except for portable units where receivers and transmitters are involved with their displays.

4.4.2.4 Desktop Publishing

Desktop publishing has much in common with some of the business graphics applications, as well as animation, in that it requires capabilities for a wide range of images, colors, and various formats. However, it does not necessarily need very high processing speed or maximum resolution, although the latter may be desirable in cases where the displayed image must be very close to the final printed result. This requirement may frequently be minimized in order to reduce cost and simplify the system. In addition, although standalone workstations are very convenient, it is fully possible for a host computer that drives a number of X Windows terminals to be used in situations such as a newspaper setting where a large number of independent terminals might be too costly and inefficient. These are the reasons why the requirements listed in Table 4.8 for this application are relatively undemanding. These could apply to the newspaper setting, but might be wholly inadequate for a high-quality printing environment such as that for art books

TABLE 4.12   Desktop Publishing Systems

	Processor Requirements		Display Requirements			
Applications	Speed (MHz)	Memory (Mbytes)	Resolution	Size (in.)	Color (No.)	System Requirements
Complex graphics	66–100	120–480	1280 × 1024	21–27	256–16 million	High-end workstation
Multiple terminals	25–30	60–80	800 × 600	17–21	256–4096	Server/X terminal workstation
Simple A/N	20–25	40–60	640 × 480	15–17	16–256	Server/ASCII terminal

or volumes that contain a large number of complex illustrations. It should be noted that these branches of the general application group must have the ability to view, modify, and lay out complex pages; hence the maximum capability may be required for these examples. Table 4.12 covers three sets of requirements, and shows the system most appropriate for each.

Desktop publishing may be described as involving all operations that must be performed in order to generate the material required to produce the final hard-copy form of the book, magazine, or other document that is desired. The necessary operation may be carried out on workstations designed for this particular activity, but modern general-purpose desktop computers, in conjunction with a variety of available software, are quite adequate for most of the operations required for the majority of publishing needs. Of course, this general-purpose equipment may have more capabilities than are necessary for some specific needs, but the difference in cost between equipment that is barely adequate and that which is beyond the immediate needs is usually so small as to justify the better system. This is true of the display portions as well as the processing sections, and the display parameter values shown in Table 4.13 reflect this choice. However, it is also possible to tailor the

TABLE 4.13   Desktop Publishing and Document Preparation

		Technology	
Parameter	Value	Input	Output
Luminance (cd/m^2)	25–50	NA	CRT, FPD
Contrast ratio (no.)	5–10	NA	NA
Resolution (TV lines)	480–1280	NA	CRT, FPD
Speed (in./s)	2–20	CRT, FPD	CRT, FPD
Colors (no.)	256–16 million	CRT	CRT, FPD
Viewing area ($H \times W$, in.)	7 × 10–15 × 20	NA	CRT, FPD
Matrix (columns × rows)	512^2–1280^2	NA	FPD
Matrix element (in.)	0.01–0.001	NA	FPD

requirements more closely to the specific needs, and this choice is covered by the lower values. Again, it should be noted that this type of choice is probably not economically warranted, but the decision may be made to follow this route, if desired. The final decision in this and other applications is left to the user. These requirements are similar to those for process control and monitoring, but differ in the number of colors and maximum resolution. This is because the process control and monitoring functions have less stringent demands for total data and legibility. In addition, there is no need for a large-board display. Document preparation in general has a less rigorous set of parameter requirements. This may not be true for certain in-house publications, but in these cases one may consider the preparation of these documents as part of desktop publishing, and the requirements will be the same.

4.4.2.5 Education
Education is an application of increasing importance as computers become as commonplace in the schools as in the home. Indeed, electronic games, covered separately in Chapter 8, are also used as part of the educational arsenal of teaching aids. The requirements can cover a wide range depending on whether the display is used to provide only A/N information, or to illustrate some complex process using detailed visual images. It has been demonstrated that children at the age of 5 are capable of using computer-generated images and operating systems (LOGO) to construct their own learning material; thus computers are being used to an increasing extent even in the lower levels of grade schools. This is a long way from the original intentions of Bitzer and Slottow when they first developed their learning models and systems, but it is surely the wave of the future. Therefore, education as an application warrants considerable attention and separate treatment in some detail. Indeed, education may become even more important with the recent trend to provide Internet access at most of the public schools. A set of parameter requirements is listed in Table 4.14, which should be applicable to the Internet access capability.

4.4.2.6 Information Systems
Information systems as an application tend to require systems that fall into a category that does not impose severe requirements on either the data processing or display portions of the operating system. This assumes that the applicable information system is restricted to text and fixed images, and that any other type of information is included under one of the other data processing groups. With this proviso, it may be concluded that the simplest types of system, that is, a single server driving a group of ASCII terminals, is adequate for examples of this general application, although there may be a requirement for a rather larger memory and good visual resolution.

Geographic information systems (GIS) are so representative in terms of the display requirements that they may be used as the designation for the

TABLE 4.14   Education

Parameter	Value	Technology	
		Input	Output
Luminance (cd/m^2)	25–50	NA	CRT, FPD
Contrast ratio (no.)	5–10	NA	NA
Resolution (TV lines)	480–1280	NA	CRT, FPD
Speed (in./s)	1–10	CRT, FPD	CRT, FPD
Colors (no.)	256–16 million	CRT	CRT, FPD
Viewing area ($H \times W$, in.)	7×10–15×20	NA	CRT, FPD
Matrix (columns × rows)	512^2–1280^2	NA	FPD
Matrix element (in.)	0.01–0.001	NA	FPD

information systems application. GIS has mapmaking as an important part of its output and might go by that cognomen, but GIS is a more inclusive term and expands the application to contain other types of information such as temperature gradients, wind patterns, and weather charts. It might also include other charts of a nature similar to these, but not specifically related to geography, but these do not change the display requirements to any significant degree. Mapmaking is a demanding application and should be adequate to define the display requirements that are listed in Table 4.15.

4.4.2.7 Measurements and Instruments

The instruments category is combined, for the purposes of this discussion, with the measurements application, because one always implies the other; that is, instruments are always needed to make measurements, and measurements imply the existence of instruments equipped to make them. Although instruments and measurements might be treated separately, the results would be quite redundant, as the display capabilities of the various types of instruments imply that they are adequate for the measurements to be made.

There are a large number of instruments available, used for a variety of measurements. The instruments range from the simplest type such as current

TABLE 4.15   Geographic Information Systems (GIS)

Parameter	Value
Luminance (cd/m^2)	35–70
Contrast ratio (no.)	10–20
Resolution (TV lines)	480–1280
Speed (in./s)	1–10
Colors (no.)	4–16
Viewing area ($H \times W$, in.)	7×10–15×20
Matrix (columns × rows)	512^2–1280^2
Matrix element (in.)	0.01–0.001

meters, to complex analytic systems that measure many different parameters to a high level of accuracy. Most of the simpler instruments are limited to single or a few related measurements, and the requirements are correspondingly simpler, frequently restricted to numerics only or numerics with a few alphas. Multimeters that measure a few parameters such as resistance, current, and voltage are examples of this type of instrument, and the display may be an analog meter, although all-digital units are more common. Next in complexity comes a median level, where the instruments are required to present more data but only more A/Ns are needed. However, this increase leads to the use of larger panels so that the complexity of the display is increased and usually requires some type of scanning. Next we come to those applications such as analyzers and generators where the data are displayed in the form of various types of curves, and some type of oscilloscope, $X-Y$ vector, or raster display monitor is incorporated in the instruments to present these data. The raster types are covered in detail in Chapter 3, and only the data on the oscilloscope and $X-Y$ vector monitor groups are presented here. Although these groups are, strictly speaking, not in the purview of this text as separate entities, the data shown are relevant in relation to equipment that contains these types of displays as the main form of data presentation.

Finally, we come to the wide range of test equipment that use some type of monitor, either in conjunction with or as part of a computer-based subsystem. These are growing importance and are therefore treated in some detail below. They may be part of a large-scale test instrumentation system, which is coming more into vogue, but the discussion is limited to the application requirements to the display system used.

Beginning with the simplest group, consisting of the many different types of meters used to measure a host of parameters such as resistance, voltage, current, power, and many others, these usually contain only minimal readouts, generally restricted to one or two lines of numerics with a limited number of alphas. Representative examples are the digital multimeters that measure several different parameters such as voltage, current, and resistance on a single readout. For these, the display requirements are usually restricted to numerics, as is the case for the majority of panel meters. However, other meters in this group may also require simple alphas and some special characters, and the display requirement should include this, as is shown in Table 4.16a. The numerics may be formed from figure eights, but matrix formats are also used. The next group also contains characters of only the A/N type, but the display is expanded to A/N panels that may contain up to 40 or more characters in as many as four lines of characters. There may also be an increase in the number of colors as is shown in Table 4.16b. The other parameter values are similar to those for the readout types, but are listed in that table for ease of use. In addition, it should be noted that larger values are possible for all these parameters as this display is flexible in its requirements, depending on the actual equipment size and capabilities. The third category of instrument uses oscilloscopes or waveform monitor–vectorscopes

TABLE 4.16   Instrumentation

Parameter	Value
a. Readouts	
Luminance (cd/m^2)	30–70
Contrast ratio (no.)	10–20
Character lines (no.)	1–2
Characters (no.)	10–20
Character size (H, in.)	0.1–2
Response time (characters/s)	10–20
Colors (no.)	4–16
Viewing area ($H \times W$, in.)	0.5×2–2×10
Matrix (columns $\times$ rows)	5×7–7×9
Matrix element (in.)	0.05–0.5
b. Panel	
Luminance (cd/m^2)	40–90
Contrast ratio (no.)	10–20
Character lines (no.)	1–4
Characters (no.)	10–40
Character size (H, in.)	0.1–2
Response time (characters, s^{-1})	10–20
Colors (no.)	16–256
Viewing area ($H \times W$, in.)	2×10–4×20
Matrix (columns $\times$ rows)	5×7–7×9
Matrix element (in.)	0.05–0.5
c. Oscilloscope Display	
Gain accuracy (% of full scale)	± 1.5–3
Bandwidth (Hz-MHz, -3 dB)	Dc–500
Rise time (ns)	0.35 bandwidth $^{-1}$
Vertical sensitivity (mV/division–V/division)	0.5–5
Vertical resolution (% of full scale)	0.1–0.4
Sweep accuracy (%)	0.005–0.01
Sweep resolution (ps)	20–100
Sweep speed (ns/d-division–s/division)	2–5
Number of channels	1–8
Phosphor	P-31, RGB
d. X–Y Vector Monitors	
Luminance (cd/m^2)	75–100
Spot size (mm)	0.25–0.28
Bandwidth (MHz)	Dc–4
Phosphor	P-31, others
Sensitivity (mV/division)	50–150
Linearity (%)	3–5

as the main display means. Oscilloscopes are generally categorized in terms of vertical accuracy, bandwidth, and rise time, and sweep speed and resolution insofar as the display portion in concerned. These are shown in parts *c* and *d* of Table 4.16 for representative units used in the generators and analyzers that have such displays. The technology is still largely CRT, but EL, plasma, and LCD may make some inroads. Indeed, LCDs have become more common for oscilloscopes, and the other FPDs should follow with the introduction of large-screen and color units. The input and output technologies are shown only once, and may be assumed to apply in all parameters. These values are merely representative, and units with wider ranges are also available, including those that offer an RGB color capability. In addition, when incorporated into the various types of instruments that use oscilloscopes, or monitors of either the vector or raster type, the capabilities may include color and other parameters. This whole group of instruments is moving rapidly in the direction of offering a wide range of display options.

Finally, as noted previously, there is the group of measuring instruments that use computers and their accompanying raster monitors as basic components of the total measurement and test capabilities. They are frequently in the form of computer-based systems and as such are covered as part of the computer group.

4.4.2.8 *Manufacturing*
Manufacturing operations that are at least partially automated require displays that can provide information on the state of the operation. These may range from the simplest numeric to complex presentations that are used to represent each state of the total operation. Another approach is the use of simulation to create case studies of manufacturing that use supercomputers and complex imagery to reduce the numbers of trials and improve quality. These aspects of manufacturing lead to the display requirements shown in Table 4.17.

TABLE 4.17 Manufacturing Application

Parameter	Value
Contrast ratio (no.)	5–15
Resolution (TV lines)	480–1024
Speed (in./s)	5–10
Colors (no.)	4–16
Viewing area ($H \times W$, ft.)	5×10–10×20
Viewing area ($H \times W$, in.)	7×10–15×20
Matrix (columns × rows)	512^2–1024^2
Matrix element (in.)	0.01–0.005

4.4.2.9 Military Applications

The military applications have been among the most significant for electronic displays in that they imposed the most rigorous set of parameter requirements for the display systems and devices. However, this application area has become somewhat less important with the downsizing of the military establishment, and the rapid growth of many other application areas. However, the military continues to support some important display system and device developments, and also remains as an important market for electronic displays. However, this market is rather broad and diversified, corresponding to a number of the specific applications shown in Table 4.8. The particular applications that make up the military market are shown in Table 4.18.

It is not practical to cover each application separately, especially since each is treated separately in terms of its nonmilitary applications. In addition, the strong trend in military procurements is to use the closest equivalent to the nonmilitary types that will meet military environmental requirements or are militarized versions. In these cases, the display parameters are the same as for the civilian versions, and information about the display requirements for the military equivalent applications for these may be found in the sections devoted to the nonmilitary equivalents. However, one that does warrant some further discussion is avionics, which is treated separately next.

Avionics The descriptions of the HUD, HDD, and HMD are fully applicable to the military versions with the addition of the more stringent environmental and life requirements. However, the HUD versions are more advanced, with the addition of ones that use holographic techniques to generate 3D displays. A design of this type with a two-color capability has been developed, and some of its operating characteristics are shown in Table 4.19. The two-color capability is achieved by the use of the Tektronix LCD color switch, and the basic system configuration is shown in Figure 4.19.

TABLE 4.18   Military Applications

Application	Technology	Display Types
Avionics	CRT, LCD	CRT, FPD, projection
Communication	CRT, LED, LCD	CRT, FPD, readout
Control and monitoring	CRT, LCD, EL, plasma	CRT, FPD
Desktop publishing	CRT, LCD, EL, plasma	CRT, FPD, projection
Education	CRT, LCD	CRT, FPD, projection
Geographic	CRT, EL, plasma	CRT, FPD, projection
Information systems	CRT, LCD	CRT, FPD, projection
Measurements	CRT, LCD, vacuum-fluorescent	CRT, FPD, readout
Navigation	CRT, LCD, vacuum-fluorescent	CRT, FPD, readout
Presentations	CRT, LCD	CRT, FPD, projection
Transportation	CRT, LCD, electromagnetic	CRT, FPD, large-board

TABLE 4.19   Avionics Application Display Specification: Holographic HUD

Parameter	Value	Technology	
		Input	Output
Green luminance (cd/m^2)	1200	CRT	Holographic
Red luminance (cd/m^2)	750	CRT	Holographic
Switching time (us)	100–1000	LCD	LCD
Useful area (diameter mm)	73	CRT	LCD

Another development of interest is of an HMD using a specially designed hybrid chip to provide the video drive for a CRT HMD showing either stroke- or raster-generated symbology. This allows either format to be used so that the advantages of each may be obtained. The parameters for this unit are given in Table 4.20a. Both the holographic HUD and the CRT HMD systems are also applicable to the nonmilitary avionics area and should be in use there once they have achieved sufficient capabilities to become generally available for this type of usage. It should also be noted that the use of the holographic HUD has significant applications to the automotive market as well, and developments are being conducted by a number of the manufacturers of these types of transportation vehicles. This is in addition to the nonmilitary applications to commercial aircraft. Another example of an advanced avionics display for military applications is the color display that uses LCD projection technology. This is a direct-view display unit that has three active-matrix LCD panels and a xenon lamp to achieve a CRT type of

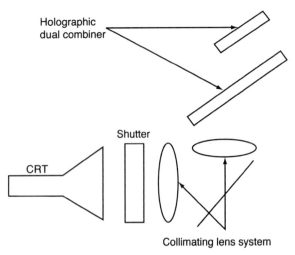

Figure 4.19   HUD architecture schematic drawing. After Ebert [11], by permission of SID.

TABLE 4.20   Additional Display Specification

Parameter	Value
a. CRT HMD	
Resolution (TV lines)	1280×1024
Video bandwidth (MHz)	100
Luminance (cd/m^2)	3000–6000
b. LCD Projection Display	
Screen size (in.)	6.0×6.0
Resolution (pixels)	512×512
Luminance (cd/m^2)	1–1500
Luminance uniformity (%)	± 10
Contrast ratio (white)	38: 1

display with the performance characteristics shown in Table 4.20*b*. This is a cockpit display and must be viewable in full sunlight, and has the potential of replacing CRT types that are presently in use in fighter aircraft and helicopters. It has the further potential of taking over a number of other avionic military applications. Significant nonmilitary avionic applications are also within reach, as demonstrated by the performance capabilities listed in Table 4.20*b*.

4.4.2.10   Process Control

This application comes under the general categories of control and monitoring but is more specific in that it is limited to one type of operation, although the actual products involved may differ widely. However, regardless of whether this product is gasoline from a refinery or drugs from a pharmaceutical plant, the operations have many points in common and the type of control and display system is essentially the same for all. The requirements are not too complex as the steps involved are usually quite limited from a control point of view, and the automaticity of the activity minimizes the number of dynamic displays needed, and the data density can be small. Thus, a server/ASCII terminal system is usually adequate for the operation, and the requirements are those shown in Table 4.4.

4.4.2.11   Simulation

This application potentially can impose the most severe requirements on the system, depending on how complex the simulation operation is. Thus, for pilot training where a close similarity to the actual situation is essential, and where changes must be in real time, either a high-level standalone workstation, or even the combination of a host data processing system and a high-level workstation may be required. It might be noted as well that electronic games impose a similar requirement, although the results of failure

TABLE 4.21   Simulation Requirements

| Application | Processor Requirements | | Display Requirements | | | Systems Requirements |
	Speed (MHz)	Memory (Mbytes)	Resolution	Size (in.)	Colors (no.)	
Complex	66–100	120–540	1280 × 1024	21–27	256–16 million	Server/high-end workstation
Moderate	30–66	60–120	800 × 600	17–21	256–4096	Workstation
Simple	20–60	40–60	640 × 480	15–17	16–256	Server/X terminal

are not as drastic as for the previously mentioned simulation. In any event, the requirements for simulation can range from the relatively simple to the highly complex, as indicated by Table 4.21.

4.4.2.12 Conclusions

It should be apparent from the analyses and data presented in this section that a broad range of system requirements may be involved in satisfying the needs of the applications covered in this chapter. The products involved range from the simplest terminal to the most complex combination of host data processing and server units with terminals and standalone high-level workstations. These systems employ many different technologies, and from the display point of view, the terminals that may be used embody all the display technologies covered in Chapter 2. The applications covered in the following chapters also employ the types of systems covered in this section. Therefore, we may anticipate that the further discussions in those chapters elaborate on the material presented previously with specific references to the requirements of those applications. However, the general equipment and system data presented here are adapted to the specific needs of the applications covered in the following chapters, and they should be read with reference to that data.

REFERENCES

1. Sherr, S., *Electronic Displays*, 2nd ed., Wiley, New York, 1993, p. 343.
2. Sherr, S., *Electronic Displays*, 2nd ed., Wiley, New York, 1993, p. 345.
3. Breen, P. T., "Review of Workstation Technology," *SID Sem. Lect. Notes* 1991, M9/11.
4. Wilson, D., "X Windows Terminals Search for a Single Processor Solution," *Comput. Design* Aug. 1991, 78.

5. Sherr, S., *Electronic Displays*, 2nd ed., Wiley, New York, 1993, p. 527.

6. Goy, C., "Mice," in S. Sherr, Ed., *Input Devices*, Academic Press, New York, 1988.

7. Adam, G., *Tactical Cockpit Systems*, McDonnell Douglas Corporation, Figure 24.

8. Sherr, S., *Electronic Displays*, lst ed., Wiley, New York, 1979, p. 565.

9. Perbet, J. N., and Parus, R., "HUD for Civilian Aircraft," *SID 1994 Dig.* 97–100.

10. Adam, G., *Tactical Cockpit System*, McDonnell Douglas Corporation, Figure 155.

11. Ebert, J. C., "Color Holographic HUD," *SID 1993 Dig.* 443–436.

PART III

USER APPLICATIONS

5

COMPUTER GRAPHICS

5.1 INTRODUCTION

Computer graphics are available in a wide variety of actual applications, ranging from advertising through animation and fine arts, to television. Some of these have been covered from a general viewpoint and are repeated here to provide a background for discussion of the specific uses of computer graphics for these applications. Therefore, in order to comprehend this range, it is necessary to describe and discuss in some detail at least a few representative applications out of the extensive list shown in Table 5.1, most of which may also be found in Table 4.8. To this relatively limited list should be added a number of other applications, covered in detail in subsequent chapters, such as visualization and virtual reality, to name some of the most advanced applications for electronic displays. However, at this point, the discussion is limited to the applications listed in Table 5.1, which are representative in terms of their display requirements. These applications are described separately in the next section, and these discussions may be considered to apply equally to the general computer graphics application. In addition, they supplement the somewhat more limited descriptions given for these applications in Section 1.3.2.

Computer graphics is a general application that includes all those listed in Table 5.1. However, it is convenient to list it separately here in order to provide a category to include certain types of outputs that would not fall directly into any of the specific subapplications. Computer graphics as a completely separate application may vanish when the use of computer-generated graphics images becomes fairly common, so basically all of its possibilities are covered as the sum of all the relevant applications. However, for the purposes of this volume, it is still appropriate to treat it as a separate application. In addition, this application is similar to the previous ones in that although a high-end workstation is the best system for the full range of requirements, it may be possible in some cases to do with the least complex

TABLE 5.1   Computer Graphics Applications

Application	Displays	Other Graphics Units
Advertising	Monitor, projector	Fixed large board
Animation	Monitor, projection	3D model, plotter, wand
Desktop publishing	Monitor	Plotter
Fine arts	Monitor, plotter	Board, sketchpad
Geographic information systems	Monitor, projection	Large board, plotter
Simulation	Monitor, projection	Plotter
Sports	Monitor, projection	Camera
Television	Monitor, projection	Camera

TABLE 5.2   Computer Graphics

Application	Processor Requirements		Display Requirements			System Requirements
	Speed (MHz)	Memory (Mbytes)	Resolution	Size (in.)	Color (no.)	
Full graphics	66–100	120–480	1280 × 1024	21–27	256–16 million	High-end workstation
Moderate graphics	25–30	60–80	800 × 600	17–21	256–4096	Server/X terminal, workstation
Simple images	20–25	40–60	640 × 480	15–17	16–256	Server/ASCII terminal

versions of standalone workstations if the actual applications are restricted to the least complex graphics. Under these conditions it is appropriate to divide the operational requirements into three groups and establish separate system requirements for each. A possible division is into simple images, those with moderate complexity, and those with maximum graphics requirements. This is done in Table 5.2, and provides an overall view of the computer graphics requirements. This is followed in the subsequent section by the description and requirements for all the subapplications listed in Table 5.1.

5.2   APPLICATION DESCRIPTIONS

5.2.1   Advertising

A short description of the advertising application may be found in Section 1.3.2.1, and it may serve as an introduction to the material presented here. Advertising is a ubiquitous application, and the displays that are used to present the advertising messages to the public are many and varied. Everyone is familiar with the fixed message billboards found in all highways, and to these should be added the electronic signs using LEDs, lightbulbs, and rotating disks that may be used to generate A/Ns, images, and other forms of fixed and variable displays. These come under the category of board displays, but in some instances projection techniques may be used as well. Next there is the broad category of displays used as output devices for TV programming, and here the display requirements are necessarily the same as those for any other TV presentation. Television as a general application is discussed further in Chapter 8, on entertainment. Suffice it to state, at this point, that the advertising message produced for TV presentation may be among the most demanding of all TV display requirements. They may include fixed images, animation, and live activities, leading to the need for high-quality displays in order to fully achieve the desired effect. Undoubtedly, HDTV will lend a helping hand with its improved image display capabilities, and add as much or more to this application as to the more general entertainment application. Even the simplest example of an advertising display would demonstrate the potential complexity of the display imagery, and the high performance level that may be required. Advertising also falls into other general application categories, of which virtual reality and television are probably the most important, and is discussed further in Chapters 8 and 10. Suffice it to state at this point that advertising may have a stringent graphics art imaging requirement.

5.2.2 Animation

The animation application is briefly described in Section 1.3.2.2, but the actual procedures employed are not discussed there. Animation has become a fully computer-generated operation, with the need for large memories that contain an extensive group of images and their animation sequences. Thus, mouth movements for speech can be generated by selecting the appropriate static image and then using the set of shapes that correspond to the desired mouth movements to complete the speech sequence. Alternatively, the initial shape may be drawn using the computer program and the sequence created by applying the steps called out by the animation program. In either case, a large number of images is required with the associated need for large memories referred to previously. This operation should be performed at a rapid rate to allow the animated images to present the appearance of the final set. The result is that a rapid data processing capability is an essential part of the animation system. These requirements lead to the choice of either a client–server X-Windows terminal combination where a number of terminals may be handled by the client–server data processing unit, or a group of standalone, high-level workstations for a number of users. The lower speed and memory requirements shown in Table 5.2 corresponds to the latter situation, and the high-resolution, multiple color display requirements apply to either choice. Again, the choices are preliminary and somewhat tentative.

One procedure in use for producing animation graphics is to begin with modeling, and then proceed through rendering and animation. The modeling step creates the graphics art image in 2D or 3D form by drawing directly on a graphics tablet or drawing the images on paper and then scanning them into the workstation where the other steps are performed. In addition, some sources provide galleries of graphics images that may be stored in the workstation and called up as desired, thus avoiding the drawing step. In addition, complicated mathematical techniques may be used to generate complex 3D graphics art images as the starting point for the animation process. The graphics images range from simple scenic views to complex panoramas, and the demands on the display system may be quite rigorous to allow the most complex images to be adequately shown. Suffice it to say at this point that although the graphics art images used for animation sequences may not achieve the level of fine art, they do exhibit all the visual characteristics generally found in computer graphics art, and should be included as one of the display applications. CRT is still the preferred technology, in the lead because of its ability to provide high resolution, adequate screen size, and full color gamut. With the improvements in FPDs, in particular that promised by plasma and FETs, CRTs may be replaced in the future. However, in the near future it seems most likely that CRT monitors will continue to prevail.

5.2.3   Desktop Publishing

A short, preliminary description of the display imaging requirement for this application is given in Section 1.3.2.14, but does not deal with the graphics art aspects that may be involved other than to note that they may exist. At this point it is appropriate to discuss these aspects in some detail to illustrate what they may be, and what types of images are involved. In particular, it is of interest to note to what extent graphics art images may be used in conducting this application and providing the visual images that may be required. To this end, a short discussion of the desktop publishing process may be useful.

Desktop publishing has much in common with some of the business graphics applications, as well as animation in that it requires the capabilities for a wide range of images, many colors, as well as various formats. However, it does not necessarily need very high speeds or maximum resolution, although the latter may be desirable in cases where the displayed image must be very close to the final printed result. This requirement may frequently be minimized in order to reduce cost and simplify the system. In addition, although standalone workstations are very convenient, it is fully possible for a host computer that drives a number of X Windows terminals to be used in situations such as a newspaper setting where a large number of independent terminals might be too costly and inefficient. This could apply to the newspaper setting, but might be wholly inadequate for a high-quality printing environment such as that for art books or volumes that contain a large number of complex illustrations. It should be noted that these branches of the general application group must have the ability to view, modify, and lay out complex pages so that the maximum capability may be required for these examples. Desktop publishing is a version of large-scale publishing that can be performed on the small but powerful PCs that are available in increasing numbers, and at constantly decreasing prices. It is a broad application, and it may include a wide range of actual products, ranging from simple brochures to magazines and books. In addition, the operational results may be restricted to in-house publication, or expand to full layouts for books and magazines. Thus, the display graphics requirement can be quite simple, or highly complex, depending on what the actual output requirements might be. It is an application with a broad potential, as it may take over the operations now performed by full-scale printing and publishing establishments. The actual operations involved may be described as the full sequence of steps required to turn out the final hard-copy version of any, if not all, of the possible publication types. These operations are usually performed on workstations that have been specially designed for this purpose, but they may also be performed on general-purpose PCs, and a good deal of software is available to perform these functions on such equipment. The actual computer graphics aspect of these desktop publishing operations consists primarily of creating the sometimes quite complex graphics images that are included in the publications. Again, these images may range in complexity from simple line drawings to elaborate illustrations that must be produced in full color and with maximum available resolution. As an adjunct to this range of possible requirements, the display portion of the equipment can be either a simple monitor, a highly complex workstation, or a system in between these extremes, all of which may use both monitors and printer–plotters. In any case, the displays are a very important part of a publishing system, desktop or otherwise. Therefore, Table 5.3 covers three sets of requirements, and shows the system most appropriate for each.

TABLE 5.3   Desktop Publishing Systems

Application	Processor Requirements		Display Requirements			System Requirements
	Speed (MHz)	Memory (Mbytes)	Resolution	Size (in.)	Color (no.)	
Complex graphics	66–100	120–480	1280 × 1024	21–27	256–16 million	High-end workstation
Multiple terminals	25–30	60–80	800 × 600	17–21	256–4096	Server/X terminal, workstation
Simple A/N	20–25	40–60	640 × 480	15–17	16–256	Server/ASCII terminal

5.2.4   Fine Art

Fine art as an application for electronic displays is discussed under the general category of art, but this term rather ignores the fact that graphics art is characteristic of a number of applications, of which fine arts is only one example, albeit of a rather unique nature. Although this application of graphics art is of a rather extensive type, it is restricted, at least at present, to the unique results of the efforts of a relatively small group of individual graphics artists. This small number of visual artists consists of individuals who are devoting much of their artistic creative efforts to using electronic displays and systems to produce their art. Thus, as an application it does not appear to attract a very large market for these displays and systems. However, the results of these efforts are of sufficient interest to warrant paying some attention to the display requirements, especially as they tend to be among the most demanding of all electronic display applications. In particular, the graphics arts aspect is most apparent in this application, and it is in some ways the best example of how graphics art is an important member of the list of applications for electronic displays. Much of the fine art produced by these electronic means tends to be of the abstract variety, but it is also possible to produce images for other purposes that have most of the characteristics of fine arts examples. In any event, the techniques involved in graphics arts production are fully applicable to the fine arts category, and it may be anticipated that their use for that purpose by serious artists will continue to grow.

As noted previously, the fine arts examples impose one of the most severe requirements on the display system, which makes this application of particular interest in any examination of what might be needed to meet the general requirements of graphics arts application. This means that large graphics tablets and large-screen monitors with full-color capability and maximum resolution are essential. In addition, high-quality color plotters are an essential component of any system used for fine art viewing and reproduction. In addition, in some cases projection systems with similar performance capabilities may be desirable, to enable the results to be viewed in something approaching final form before they are reduced to hard copy. It should also be remembered that the main purpose of the artist, when producing fine art, is the expression, and to some extent the communication, of certain attitudes toward life and experience. Of course, this initial impulse and need are often made subordinate to other principles, such as critical acceptance or financial rewards. However, the major driving force remains the individual need for self-expression and not primarily doing the job to obtain some kind of remuneration for the effort or possibly some modicum of satisfaction from that effort that may be similar to the satisfaction achieved by the artist. This makes the results of the fine art effort much more susceptible to deficiencies in the electronic display equipment and system, with the concomitant dissatisfaction with the available means. Thus, this application is probably the most demanding not only because of the type of resultant image but also because this result must fully satisfy the creator.

5.2.5   Geographic Information Systems (GIS)

The role of computer graphics in GIS may be quite limited or very extensive, depending on the type of cartography involved, and the role to be played by the final product. It should be noted that mapmaking or cartography is probably the most significant aspect of GIS insofar as the display application is concerned, and a similar situation exists for the role of computer graphics. Of course, the simpler the map, the less the impact on the computer graphics requirements. However, even a relatively simple map may have an extensive need for high performance on the visual displays of both the monitors and hard-copy devices used. The simplest map may require relatively high quality, and an even higher performance demand can result from the preparation of maps containing various illustration and drawings. As noted previously (Section 1.3.2.19), the use of color may range from as few as 5 to as many as the full gamut requires. In either case, color is important, although not completely necessary for the simplest images.

Again, the primary equipment performance requirements seem to be high resolution and multiple color capabilities, as are large size and low distortion. It may also be advisable to include projection systems to fully show the graphics art aspects of the visual image in their best representation. Finally, the ultimate output is the hard-copy representation that constitutes the final cartographic result, so high-quality plotters are an essential part of any display system used for these purposes. Of course, the final production of the finished maps in whatever form they may be desired—from small individual types to the large, complex ones found in atlases and navigation documents —is the result of another application type: publication, whether desktop or large scale printing. The desktop application is covered in Section 5.2.3, and the printing applications are beyond the purview of this volume.

5.2.6   Simulation

Computer graphics may play a very large role in simulation, especially when the simulated environment is created in a close emulation of the real world. This is particularly true of those systems that are used for training the user to operate various types of equipment without actually exposing the operator to the actual equipment. Prime examples of these types of systems are the ones used for aircraft and vessel pilot and navigation training. In their most elaborate form the simulation is such that a recognizable equivalent of the visual environment is created and the training process maintains this condition throughout. Simulated images can achieve a close approach to reality, and the graphics art requirements for achieving this level of simulation are quite extreme, demanding the highest level of visual performance from the display equipment and system. This led initially to the use of CRT-based monitors, but the improvements in monitors using FPDs, in particular the AMLCD types, and the large plasma panels have made these FPDs acceptable substitutes for CRT monitors. However, in specific types of training systems, the need for large displays with 180° or more of coverage has made projection systems the technology of choice. CRT-based projection systems have been the preferred technology, but LCD projectors are rapidly approaching equivalent performance, and may replace the CRT types in the near future.

5.2.7 Sports

The application of computer graphics displays to sports makes up what is perhaps one of the least anticipated, but surprisingly successful examples of the effectiveness of well-designed graphics in properly chosen environments. To understand how computer graphics images can be used in sports events, it is necessary to examine what other displays are used in such situations. Until recently, the graphics capabilities available to sports broadcasters were quite limited, but advances in the technology have enabled these users to considerably expand the role of graphics in their programming. As a corollary to these advances, the computer graphics aspects have been expanded as well. More specifically, one network uses what is termed a "bold, futuristic style," consisting of using a variety of art techniques to provide visuals that combine information with images of the sports activity. This type of usage of graphics art elements to enhance the sports presentation is still somewhat in its infancy, but is expected to grow as the technology improves and the broadcast stations make the graphics technologies available to their users. Although it is still too soon to establish what the requirements might be, the fact that the graphics results are achieved by means of the same programs and equipment used for other high-quality images, including 3D graphics. Thus, it seems reasonable to state that the same monitor types are needed and will be used for this application as for the other applications requiring high-quality electronic displays using both CRT and FPD technologies.

5.2.8 Television

As a corollary to the sports application, and defining most of the computer graphics requirements for sports, there is the broad field of television. This is surely one of the most demanding application for graphics art, rivaling in many respects that established by fine arts. The requirements have been made increasingly stringent by the imminent arrival of high-definition television (HDTV) on the market in the United States. The wide variety of programming shown on TV, ranging from simple weather maps to sophisticated artwork, animation, and visual imagery for advertising purposes, defines the need for high-end electronic displays. The vast improvements in the visual outputs from CD-ROMs (compact disk–read-only memories) and the new video DVDs (digital versatile disks), coupled with the improved monitors needed for HDTV, will surely lead to an increase in the computer graphics contributions to TV programs. These contributions to the need for the highest quality of visual display lead to the requirement for high-performance monitors and projection systems.

5.3   TECHNOLOGY AND PRODUCT DESCRIPTIONS

5.3.1   Introduction

The display technologies and products found in all aspects of the computer graphics applications are quite similar, and may be discussed together for the purposes of this analysis. In addition, the technologies and products may be combined for the purposes of this section, as the technologies are directly related to the products in that any of the technologies may be used for each specific type of product. Thus, there is no loss of generality, and the particular aspects of either topic may be given adequate coverage with this approach. However, as an introductory step, it is useful to list the technologies in conjunction with the products that use them in Table 5.4.

This list is restricted to the specific display technologies and products used for computer graphics, and does not include any of the other products that may be used for total systems, such as those listed in Table 5.1. These are covered in some detail in Chapter 3, and are treated again in Sections 7.3 and 7.4 in relation to their performance capabilities as they impact on the operational requirements for multimedia and presentations applications. The discussion and descriptions contained in the following sections are supplemental to those found in Chapters 2 and 3, and are limited to those capabilities and characteristics that impact most on the general applications that are the subject of this chapter. Therefore only limited detail is included, with reference to the expanded discussion in the previous chapters when required. It is important to note that the graphics arts application has rather severe display requirements for all the subsidiary applications listed in Table

TABLE 5.4   Technologies versus Products

Product	Technology	Advantages
Flat-panel	AMLCD, PMLCD	Color, low power
	EL, FED, plasma	Luminance, size, cost
Monitor	CRT	Data density, color
	FPD	Size, power
Plotter–printer	Pen, electrostatic	Size, quality
	Inkjet, thermal	Quality, color
Projector	CRT, LCD	Size

5.1, and the technologies and products that may be used are representative of the best performance available. However, it should also be recognized that when other factors such as cost and availability are considered, it may be necessary to accept lower performance capabilities, and this is recognized in determining to what extent less-than-optimal performance may be adequate. In addition, it should be understood that the best display performance should always be the goal and lesser capabilities accepted only when this is unavoidable.

5.3.2   Dynamic Display Technologies

Most of the technologies covered in Chapters 2 and 3 and listed in Table 2.1 are involved in the electronic displays used for computer graphics applications. The one most popular is the color CRT, particularly in its implementation for CRT monitors. This has also been the case for projectors, but LCDs have begun to take over the projection function. Similar to the FPDs, the AMLCD is in the forefront for these applications, although PMLCD does show some promise. Table 5.5 contains a list of the most important parameters and the range of performance that is required for any aspect of the computer graphics applications.

The operating parameters of these units are in a constant flux, and the values shown here are only representative and not necessarily the best. In addition, several other technologies such as EL and FED are developing rapidly, and may soon be in a position to provide units with equivalent or better performance. In any event, there is a plethora of choices for all these applications, and there should be no difficulty in meeting performance requirements.

5.3.3   Hard-Copy Technologies

The prime hard-copy technologies in use for these applications are those associated with the production of physical prints, plots, and images that can be made available to the viewer for further study and take-home purposes. These may be quite elaborate, especially for the fine arts example, and of course, slides and film are also forms in which these types of items may be produced, and the hard-copy technologies involved for the production should not be ignored. The technologies are quite similar for both application groups and are combined for the purposes of analysis of the parameters as shown in Table 5.6. These examples of hard-copy output display products are by no means inclusive of all possibilities, but they do represent the main types that are available and are used for both multimedia and presentations applications. There are also some very high-quality, specialized units that use dye transfer in a fashion similar to that used for thermal transfer, but these are rather expensive and are found only in specialized applications. However,

TABLE 5.5   Technology Parameters

Technology	Parameter	Range
CRT	Display size (diagonal in.)	17–31
	Resolution (H × V)	640 × 480–1600 × 1280
	Colors (no.)	256–16 million
	Response time (μs)	1–10
	Contrast ratio	100
AMLCD	Display size (diagonal in.)	9.6–13.1
	Resolution (H × V)	640 × 480–800 × 600
	Colors (no.)	256–256,000
	Response time (ms)	80
	Contrast ratio	60
PMLCD	Display size (diagonal in.)	9.6–11.3
	Resolution (H × V)	640 × 480–800 × 600
	Colors (no.)	256–256,000
	Response time (ms)	300
	Contrast ratio	18
EL	Display size (diagonal in.)	15–40
	Resolution (H × V)	1280 × 1024
	Colors (no.)	16–260,000
	Response time (ms)	0.1–1
	Contrast ratio	10–40
CRT projection	Display size (diagonal in.)	60–240
	Resolution (H × V)	1280 × 1024
	Colors (no.)	16 million
	Contrast ratio	100
LCD projection	Panel area (in.)	7.9 × 5.3
	Resolution (H × V)	720 × 480
	Colors (no.)	2700
Plasma panel	Panel diameter (in.)	26, 40, 42
	Resolution (H × V)	640 × 480, 850 × 480
	Colors (no.)	260,000–16.7 million
	Luminance (lux)	300

the types covered, in conjunction with the dynamic display units described previously, make up an adequate group of display types.

5.4   SPECIFICATIONS AND BLOCK DIAGRAMS

5.4.1   Specifications

5.4.1.1   Introduction

The specifications described and analyzed here are for systems that are adequate for the entire group of applications that fall into the computer graphics category. They may be used to define the performance requirements

TABLE 5.6   Hard-Copy Technology Parameters

Technology	Parameter	Range
Pen plotter	Display area (in.)	12 × 12–30 × 40
	Resolution (dot size in.)	0.001–0.005
	Pens (no.)	1–8
	Colors (no.)	1–8
	Speed (in./s)	1–20
	Accuracy (% of movement)	0.01–0.05
Electrostatic	Display width (in.)	24–48
	Resolution (dpi)	200, 400
	Colors (no.)	4–16
	Speed (ips)	0.3–2.5
	Accuracy (% of line)	0.1–0.2
Inkjet	Resolution (dpi)	180–360
	Speed (cps)	167
	Jets (no.)	12–64
	Colors (no.)	256–16 million
Laser	Resolution (dpi)	240, 300, 400, 600
	Text speed (ppm)	4–120
	Graphics speed (gppm)	0.5–5
	Colors (no.)	16–256
Thermal transfer	Resolution (dpi)	200–300
	Speed (ppm)	1–3
	Colors (no.)	16–16 million
	Size (in.)	8.5 × 11–11 × 17
Slide production	Resolution (TV lines)	800–8000
	Colors (no.)	16–16 million
	Speed	30 s–5 min.

for this specific category, and as the data source for the preparations of operational requirements for the hardware and software. In particular, the performance capabilities of the dynamic and hard-copy displays are of prime interest at this point. These are covered in a general way in Tables 5.5 and 5.6 and are developed for specific examples in the applications for graphics arts as shown in Table 5.1. Several examples are selected from those shown in these tables, and representative specifications presented for the displays used in these applications, with appropriate discussions of how the specifications are chosen from the possible performance range. Next, to supplement the information contained in the specifications, block diagrams of the systems involved in producing the output displays are given so that the general form of the implementation may be reviewed and described. These descriptions include the most significant characteristics of the hardware, software, and operational procedures involved in producing and using the systems needed to meet the operational requirements of the specific applications. The specifications may be quite similar for all of the applications under discussion, but it is still advisable to provide separate tables for each one, if only to indicate the similarities and differences, small or large, among these varied applications, all of which use computer graphics as a component of the total operation.

5.4.1.2 *Advertising*

Advertising in general, and more specifically in the TV and magazine versions, use the computer graphics capability to a very large extent. The value of a graphics arts presentation of the advertising visuals appears to be of great significance to the advertising agencies, judging from the great attention that is paid to the detail of the image, either in the dynamic versions shown in TV presentations, or in the hard-copy types used in magazine illustrations. To achieve the maximum effectiveness, it is necessary for the visual specifications to call for the highest quality. This is achieved by the use of the most effective hard-copy devices and technologies so that the actual appearance of the final printed copy may be determined from the check output prints. In addition, the dynamic display capability must be sufficient to show the potential final appearance, and CRTs are still best for this purpose. Table 5.7 defines the requirements for this type of imagery, and indicates the level of quality that may be required for the most glossy magazine publication. It is sufficient to indicate the type of imagery involved and demonstrate the minimum levels of quality that may be required. Of course, the actual advertisements that are used depend greatly on the total market expected for the product that is being advertised.

The performance ranges shown in Table 5.7 do include lower levels that do not allow the display to be of the highest quality; this is because, in the absence of HDTV, the performance capabilities required by NTSC standards still apply, and cannot be improved when the transmission system adheres to these standards. However, the higher levels of performance should come into effect in the near future, and allow something much closer to the full graphics arts quality to be obtained in the standard TV system.

TABLE 5.7   **Dynamic Display Specification for Computer Graphics Advertising Application**

Parameter	Value
Luminance (cd/m^2)	30–300
Contrast ratio (no.)	10–50
Resolution (TV lines)	480–1600
Colors (no.)	256–16.7 million
Viewing area (diagonal in.)	21–40

5.4.1.3 Animation

Animation, as an example of the computer graphics application, exists in a number of forms, with the specification requirements varying over a fairly wide range, depending on the exact type of animation involved. If this animation is of the quality generated for the Disney epics, then the need for high-quality graphics arts outputs is evident, requiring a high-resolution capability in both the dynamic display and the hard-copy unit, and leading to the specifications shown in Table 5.8 and 5.9, respectively. The hard-copy unit technology is thermal transfer, because this technology offers the best color rendition, which is a prime requirement for the graphic.

5.4.1.4 Desktop Publishing

Publishing in general, and desktop publishing in particular, are applications that are prime potential users of computer graphics as a component of the publishing operation. This is clearly evident from the types of illustrations that may be used in producing the final copy, as well as the demands made on the system to create the varieties of fonts and images involved. To achieve the desired results, the display system must meet the same level of performance as is required for some of the applications covered previously. This leads to a set of requirements for the dynamic displays as shown in Table 5.10 that should meet these requirements. In addition, it should be noted that the requirements for hard copy are similar, but not as necessary for the applications. Table 5.10 covers three sets of requirements, and shows the system most appropriate for each. In general, hard copy is not customarily required for this application. However, in the event that it is considered

TABLE 5.8 Dynamic Display Specification for Computer Graphics Animation Application

Parameter	Value
Luminance (cd/m^2)	100
Contrast ratio (no.)	20
Resolution (TV lines)	1200–1600
Colors (no.)	4000–16.7 million
Viewing area (diagonal in.)	32–50

TABLE 5.9 Hard-Copy Display Specification for Computer Graphics Animation Application

Parameter	Value
Resolution (dpi)	200–300
Speed (ppm)	1–3
Colors (no.)	256–16 million
Size (in.)	8.5×11–11×17

TABLE 5.10   Dynamic Display Specification for Computer Graphics Desktop Publishing Application

Processor Requirements Parameter	Display Requirements Value		
	Complex Graphics	Multiple Terminals	Simple A/N
Luminance (cd/m²)	100–300	50–100	20–50
Contrast ratio (no.)	10–30	10–20	10–20
Resolution (TV lines)	1280 × 1024	800 × 600	640 × 480
Colors (no.)	256–16 million	256–4096	10–256
Viewing area (diagonal in.)	21–27	17–21	15–17
Speed (MHz)	66–100	25–33	20–25
Memory (Mbytes)	120–480	60–80	40–60
System (type)	High-end workstation	Server, X terminal, workstation	Server, ASCII terminal

desirable or necessary, the specifications given in Table 5.9 should be more than adequate.

5.4.1.5   Fine Art

This application is probably more honored in its neglect than its large-scale use. However, there are a sufficient number of dedicated graphics artists who use various electronic media to create visual images that they consider to be examples of fine art, as differentiated from even the aesthetically best commercial art, to make some examination of the requirements for this aspect of graphics art to be examined, at least to some extent. As noted briefly in Section 1.3.2.3, the fine arts requirement calls for the use of CRT monitors and high-performance plotters. It should be recognized that for the resultant imagery to qualify as fine art, the visual results must match that found in art produced by the more common manual means. Thus, a wide color range, the best resolution possible in both the dynamic displays and hard-copy units, and generally the highest performance levels are necessary. As a result, fine art is probably the most demanding application for graphics arts to be found, and constitutes a major challenge to designers and users. The level of this challenge is indicated in Tables 5.11 and 5.12 for dynamic and hard-copy displays, respectively. These specifications are derived to some extent from Table 1.4, but are expanded to meet the specific computer graphics fine arts requirements, and to include the hard-copy aspect of the total fine art activity. Thus, these tables should be adequate for determining the complete specification requirements. Although some of these requirements may be difficult to meet with the present level of technology, it may be assumed that they will be achieved in the relatively near future, as business

TABLE 5.11   Dynamic Display Specification for Computer Graphics Fine Art Application

Parameter	Value
Luminance (cd/m^2)	30–300
Contrast ratio (no.)	10–50
Resolution (TV lines)	1024–3600
Colors (no.)	4000–16.7 million
Viewing area (diagonal in.)	21–40

TABLE 5.12   Hard-Copy Display Specification for Computer Graphics Fine Art Application

Parameter	Value
Resolution (dpi)	200–700
Speed (ppm)	3–5
Colors (no.)	4000–16 million
Size (in.)	8.5×11–100×240

and commercial requirements increase, and costs decrease. In any event, there may be some future for the fine art application of electronic displays.

5.4.1.6   Geographic Information Systems (GISs)

The mapmaking aspect of the GIS computer graphics application is probably the most demanding aspect of the general GIS applications. This is particularly true when the various altases and other consumer-based products are considered. This is because when the general consumer is considered, the computer graphics results must be made as appealing as possible, containing a variety of high-quality images in an excellent state of reproduction. When to this consumer satisfaction requirement is added the broad category of maps used for navigation, weather information, and a host of other maplike displays, the performance requirements for the graphics arts aspect of the general application become quite important and severe. Therefore, to meet all these various demands on the displays and other equipment used to produce the satisfactory cartographic results, the performance of the electronic display portions of the systems must be high.

This high requirement is best illustrated by means of the parameter values shown in Table 5.13 for the dynamic display, and Table 5.14 for hard copy. It should be noted that both types of displays are needed to produce the graphics arts aspects for this application group, and the requirements may be more severe for the hard copy than for the dynamic display unit, as the final result is primarily the hard-copy type. The dynamic units have been largely of the color CRT monitor variety, but the AMLCD and some of the larger color

TABLE 5.13 Dynamic Display Specification for Computer Graphics GIS Application

Parameter	Value
Luminance (cd/m^2)	100–300
Contrast ratio (no.)	10–20
Resolution (TV lines)	800–1600
Colors (no.)	4000–16.7 million
Viewing area (diagonal in.)	21–225

TABLE 5.14 Hard-Copy Display Specification for Computer Graphics GIS Application

Parameter	Value
Resolution (dpi)	200–700
Speed (ppm)	1–3
Colors (no.)	4000–16 million
Size (in.)	8.5×11–21×40

plasma displays that are becoming available should also be considered. In addition, projection units may also be required to allow a full-scale examination of the final images.

The largest diameter shown for the dynamic display is for when projection is found to be desirable. Similarly, the large hard-copy maximum is for the situation where a full-scale map is to be examined in its final form. The maps involved may be quite complicated, and significant amounts of graphics art types of imagery included. Maps that contains such imagery are those that present complicated ground images and weather displays. Other examples are maps provided for travelers who wish to visit certain areas to see the arts and architecture and welcome preliminary views that may be seen in various types of publications that contain graphics art versions of these locations. The graphics art may be extensive, and constitute an important aspect of cartographic representations.

5.4.1.7 *Simulation*

The significance of computer graphics for the simulation application is noted in Section 5.2.6, and these requirements must be stressed at this point, with particular emphasis on the reality aspect. In addition, the importance of projection systems capable of providing 180° of viewing capability for certain training systems bears repetition. A similar requirement for 3D viewing also exists, and is attainable with startling realism by certain projection systems. This capability is only partially matched by smaller direct-view systems, in particular those that use CRT monitors. However, both technologies are in

TABLE 5.15 Dynamic Display Specification Computer Graphics Simulation Application

Parameter	Value
Luminance (cd/m^2)	40–300
Contrast ratio (no.)	10–20
Resolution (TV lines)	480–1600
Colors (no.)	256–16.7 million
Viewing area (diagonal in.)	21–225
Angle of view (°)	60–360

use for training and other entertainment and educational applications that benefit from the inclusion of high-quality graphics arts presentation. The Disney theme parks contain examples of the latter, and set the standard for graphics arts in simulation systems.

There is generally no need for hard copy of the graphics arts images, but if required, the specification shown in Table 5.14 should be adequate. The lower resolution shown in Table 5.15 is adequate for some of the applications, and may be met by some of the FPD monitors.

5.4.1.8 Sports

The sports computer graphics application (see Table 5.16) is still somewhat limited, although the dynamic display is extremely important for instant replay and related informational presentations. However, there is some indication that a more extensive use of graphics as part of the total electronic display usage is imminent. One example of such possible use is the representation of figures engaged in some type of sport activity, either in realistic or animated form, and expansion of this mode of display to enhance the sport presentation is certainly in the offing. The mode of such presentations is by means of the projection systems that are used for instant replay. These are primarily of the Eidophor type, and are used for live-action replays. However, it should be noted that there is also extensive use of television to show the sports activity as one of the entertainment aspects of home TV. In

TABLE 5.16 Dynamics Display Specification for Computer Graphics Sports Application

Parameter	Value
Luminance (cd/m^2)	100–300
Contrast ratio (no.)	10–20
Resolution (TV lines)	800–1600
Colors (no.)	4000–16.7 million
Viewing area (diagonal in.)	21–225

TABLE 5.17   Dynamic Display Specification for Computer Graphics Television Application

Parameter	Value
Luminance (cd/m^2)	100–300
Contrast ratio (no.)	10–20
Resolution (TV lines)	480–1600
Colors (no.)	256–16.7 million

addition, TV monitors are found in various locations at racetracks, so that dynamic displays of that type should be considered as a TV application of sports.

5.4.1.9   Television

Finally, there is the most ubiquitous application of computer graphics to electronic displays; television. Here the computer graphics aspect can be quite significant, as the images produced may range from a limited use to the extensive involvement found in some of the programs that feature museum exhibitions of fine art, as well as some of the programs that attempt to teach drawing and painting techniques. The visual displays may be limited to monitors in many cases that may range from the highest-quality color CRT types to some of the minimal FPDs used for portable receivers. However, the improvements in performance of FPDs are making them tenable, as is the case for the large-screen projection systems. This application of computer graphics requires the high-performance capabilities and viewing area range shown in Table 5.17.

5.4.2   Block Diagrams

5.4.2.1   Introduction

The block diagrams of the equipment used for the group of computer graphics applications consist primarily of the circuits involved in producing images on monitors and projection systems. However, the sources of data for these images, as well as the processing that must take place before the images can be generated, may differ considerably for the different applications. Therefore, it is useful to provide somewhat more detail in the block diagrams than might be necessary if only the display systems alone are considered. This variety of total systems is indicated by the diagram shown in Figure 5.1, which presents a generalized system for any of the applications discussed previously.

The generalized blocks shown in Figure 5.1 may be made more specific by limiting the elements to the requirements of particular applications. This is done in the following sections.

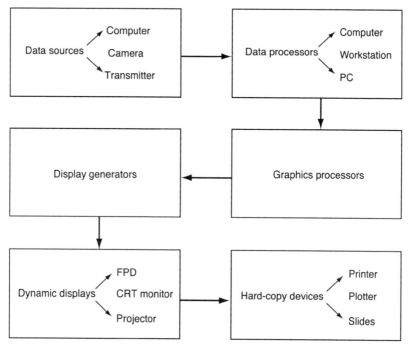

Figure 5.1   Block diagram of generalized display system.

5.4.2.2   *Computer-Based Applications*

The bulk of the computer graphics applications listed in Table 5.1 are mainly computer-based, and for these the block diagram is that shown in Figure 5.2, which is a simplified version of the block diagram shown in Figure 5.1. The major changes are in the reduction of the data sources to contain only the computer, and similarly, the limitation of the data processor types to the PC units. These changes apply primarily to advertising, animation, desktop publishing, and simulation, and to a lesser extent to fine art, geographic information systems, and television. However, the latter group differs sufficiently to warrant a separate block diagram, as is noted in the next section. For the group covered in this section, the diagram shown in Figure 5.2 may be considered adequate.

5.4.2.3   *"Other Applications" Block Diagram*

As a final example of a block diagram, that shown in Figure 5.3 may be considered appropriate for those applications not covered by Figure 5.2. It may be noted that the major difference is the addition of, or restriction to the transmitter and/or camera of the data sources. This change is particularly applicable to the television application, where the camera may be the only source when it is used on site, or a transmitter may be the only source when the display is remote and not accessible through direct connection.

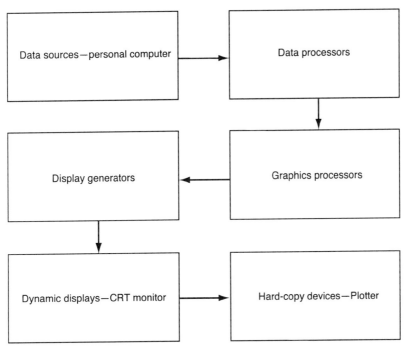

Figure 5.2   Block diagram of graphics-art-specific display system.

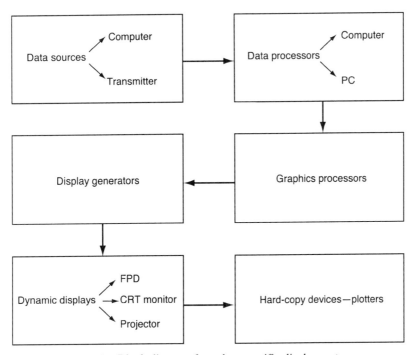

Figure 5.3   Block diagram for other specific display systems.

The similarity among these three figures is obvious, but each is shown individually to emphasize the differences and demonstrate that the basic block diagram shown in Figure 5.1 is applicable to the entire graphics arts group of applications, and the same system components may be used for all. The other block diagrams demonstrate that it is possible to simplify the equipment requirements when the system use is limited to any extent.

5.5 CONCLUSIONS

This completes the review and analysis of the computer graphics display applications, and demonstrates the extent to which computer graphics, as an application, affects the subgroups that fall into this category, as well as the specifications and equipment required to meet the operational needs of the user. Further analyses of these needs may be found in the subsequent chapters dealing with other specific applications.

6

VISUALIZATION AND IMAGING (VISIM)

6.1 INTRODUCTION

The engineering and scientific communities have long agreed on the somewhat questionable maxim, "A picture is worth a thousand words," or its opposite, "A word is worth 10^{-3} pictures." Of course, if this equivalence is calculated, using 32-bit words, then a picture with a 1024 × 1024 resolution has 1,048,576 elements that, if divided by 32, equals 32,736 words, which is considerably larger than 1000. Similarly, at 512 × 512 resolution, there are 262,144 elements, which, when divided by 32, equals 8192, which is close enough to 1000 to seem to bear out the contention. Of course, the words referred to are text rather than numbers, so that if the average number of letters in a word is assumed to be 8, then a thousand textual words equals 8000 digital words, and the claim is better substantiated. However, even if this dubious equivalence is ignored, these maxims still leave us with a sense of how pictures can expand and enhance the information content and meaning of data when presented in image form. In any event, the process of presenting data in the form of pictures or images has received the official blessings of these groups by being provided with a formal designation, namely, "visualization," as the result of a National Science Foundation (NSF) study issued in 1987, [1]. Formally, then, visualization may be defined as both the means and the activity of manipulating in an interactive fashion both large and small amounts of data, and then presenting those data in some visual form that facilitates the interpretation of those data. Some well-known forms in which such visualized data are presented are the various forms of graphs that constitute a popular means for visualization, as does animation.

A more recent application of visualization is the presentation of volumetric data in 3D visual form. This is a technique for exploring the inner aspects of volumetric objects in order to discover some of the information that may be hidden in the raw data. The data are obtained either from sampling a real or simulated model, or from some form of a geometric model. These data are obtained empirically and converted into datasets that exist in 3D discrete voxel space, where voxel is a term defining a unit volume cell, and the dataset is stored in a cubic frame buffer that is made up of a large 3D array of voxels.

Imaging is in some respects a subset of visualization in that it is also concerned with the presentation of information in visual form, although its major application is in the generation of this information in the form of documents. Thus, it may be considered as the next step after the visualization of data in dynamic form, and encompasses most of the applications included under visualization. Therefore, in the interest of conciseness, the two types of display applications are combined under the rubric of visualization–imaging, or VISIM, to coin an acronym, and the application subgroups are considered as belonging to this combined general application.

The expansion of the use of VISIM is further demonstrated by the list of applications given in Table 6.1, which have been selected from a much more extensive list, but are considered sufficiently representative and varied to serve as adequate examples. Most of the others are covered elsewhere in this volume in relation to other applications. This is still too extensive a list to permit a full description of each application in terms of its VISIM requirements within the scope of this volume. Therefore, the actual discussion is limited to CAD/CAE, documents, earth resources, geographic information systems, imaging, mathematics, medical, presentations, and scientific data. These choices are made not to indicate that the others listed in Table 6.1 are of significantly less importance, but rather to limit the discussion to reasonable lengths, with the understanding that the excluded applications have essentially similar uses for VISIM. In any event, the broad range of applications involved in VISIM when it is considered as a general application, in conjunction with the differing requirements of the viewing or user audience, lead to visualization images, presentation means, and document types that take a number of forms. This consideration leads to another point of general interest: the question of how the display requirements are affected by the types of user audiences that may be involved. These audiences may be described by the following phrases as three types:

Single Viewer. A presentation for a single individual that mainly satisfies that individual's needs and may be relatively simple, but contains much interaction capability. This may be limited to visualization but sometimes includes imaging.

TABLE 6.1   VISIM Application Areas

Aerospace	Architecture	Astronomy	CAD/CAE
Documents	Earth resources	Geographic information systems	Mathematics
Medical	Presentations	Publishing	Scientific data

TABLE 6.2   VISIM Display Equipment Types

Audience Type	Display Type	Other Units
Individual	Monitor	Audio, hard copy
Peer group	Monitor, projector	Audio, hard copy
General group	Projector	Audio, hard copy, video, VCR

Group Viewing. A presentation for a group of one's peers that is usually more visually complex and striking, that also may include some interaction. This almost always includes some imaging.

Multiple Group. A public presentation for varied groups of individuals that may include the customer or sponsor of the program that is being described. This presentation is usually quite complex, with advanced techniques in use, but with minimum interaction. Imaging may be included in the form of documents that are available for further study. The display equipment types that may be used for these three categories of types of presentation are shown in Table 6.2.

6.2   APPLICATION DESCRIPTIONS

6.2.1   Introduction

The applications discussed in this and following sections are the limited group noted previously, and selected from the larger group listed in Table 6.1. All are also covered, at least to some limited extent, in Section 1.3.2, and it is the purpose of this section to expand on those descriptions, emphasizing the VISIM aspects of the electronic display portions of the operating systems. Most notable is the transformation of A/N and other nonvisual data into some visual form that clarifies the significance of the data. The amount of complexity required in this form of data presentation is completely determined by how detailed and elaborate the original data may be. For example, when presentations are made to a large group, the visualization must be kept to the minimum complexity compatible with conveying the information contained in the original data. Under these conditions, the slides that are used and the accompanying documents must be kept simple, and the projection system used need not necessarily have maximum resolution capability. However, in general it is best to allow for a capability that may be readily expanded to produce high-quality images in the event that the electronic display or hard-copy system need to meet these more stringent requirements. These are discussed in somewhat more detail in the following sections. However, before embarking on these descriptions, it is useful to refer to the list of the display products that have been introduced in Table 6.2, in conjunction with the technologies used for each type of unit, and the

TABLE 6.3 Technologies versus Products

Product	Technology	Advantages
Flat-panel	AMLCD, PMLCD	Color, low power
	EL, FED, plasma	Luminance, size, cost
Monitor	CRT	Data density, color
	FPD	Size, power
Plotter–printer	Pen, electrostatic	Size, quality
	Inkjet, thermal	Quality, color
Projector	CRT, LCD	Size

advantages that pertain to each product and technology. This list is shown in Table 5.4, repeated here for convenience as Table 6.3.

Once again, as noted in Chapter 5 for the computer graphics, the list in Table 6.3 includes only the products and associated technologies that may be used for the generalized VISIM application. In addition, it should be noted that the use of hard copy is most appropriate for the "single viewer" audience type, listed in Table 6.2, in which the hard-copy device may become an integral part of the display system, whereas it is optional for the other audience and/or end-user situations.

6.2.2 CAD/CAE

Computer-aided design (CAD) and computer-aided engineering (CAE), are briefly described in separate sections (1.3.2.6 and 1.3.2.7), and are combined here because of the similarities in their display requirements. The main function of the visualization operation is to enable the designer or engineer, who is involved in performing any of the functions involved in producing a finished design or engineering specification, to convert the large amount of design data that might emerge from the CAD/CAE operation into visual form. CAD is used mainly to establish and define the geometry for an architectural structure, building layout, electronic circuit, mechanical part, or other similar design product. CAE then is used to analyze the CAD geometry and allows the designer to simulate and study how the product will behave so that the design can be refined and optimized. Somewhat more succinctly, it may be stated that CAD is any operation that uses computer capabilities to perform the design function, and CAE involves the use of computers to analyze the results of the design operation. In any case, any of the definitions emphasize the role of the computer or data processing facility to facilitate the design process, and arrive at viable designs by providing a full set of manufacturing instructions and drawings. The documentation in the form of these instructions and drawings is produced by adding the drafting function, which leads to computer-aided design and drafting (CADD).

All aspects of these design, engineering, and drafting operations are highly dependent on the availability of some form of output device or system, whether dynamic displays on monitors, or hard copy from printer–plotters. Thus, the electronic display capabilities play a significant role in carrying out these various steps in the total procedure from concept to product. The place of these displays in the total system is covered later, but at this point it is convenient to preface this later description with a block diagram of a system that can be used for either or both of the computer-aided operations. This is shown in Figure 6.1, which is representative of the types of equipment used.

These days, the "intelligent terminal" function is increasingly carried out by different types of personal computers, but it is also possible to use terminals that are driven by remote client server processors, as shown in Figure 6.2, so that either configuration is feasible. It might also be noted here that computer aided manufacturing (CAM) operations may also be served by similar systems, although the display requirements are somewhat less stringent.

As to the type of display equipment that may be used, CRT monitors remain in the forefront because of their high resolution and color gamut capabilities, although there are some indications that new developments in FPDs, in particular those using plasma and FET technologies may lead to FPDs and monitors that meet or even exceed those capabilities.

6.2.3   Documents

Document imaging is one of the most active application groups for imaging, and accounts for a significant proportion of the total VISIM application groups. This is particularly evident when the total revenues and numbers of user groups are considered, as is shown in Figure 6.3a, b. Another impressive statistic is the number of documents per year handled by a specific imaging system for the Sallie Mae organization, where numbers like 54 million

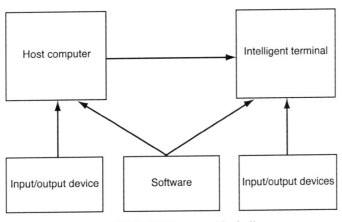

Figure 6.1   CAD/CAE system block diagram.

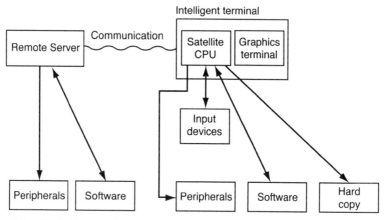

Figure 6.2 Block diagram for host–satellite CAD/CAM system configuration. After Machover [2], by permission McGraw-Hill.

records per year, stored and made available to 3000 users, attest to the magnitude of the market. Imaging systems of various sizes are offered so that small as well as very large users can be served. The block diagrams for some of these systems are presented and discussed in Section 6.4.2, but at this point the coverage is limited to pointing out the size and importance of this market, and the impact it has on requirements for monitors and image displays in general. It should also be noted, that whereas CRT monitors are still the technology of choice, the latest development in FPDs, in particular large-size, high-resolution AMLCD panels, are making FPDs a strong competitor for this and other VISIM application groups. However, it should be remembered that the requirements for document imaging may be very stringent, depending on the amount and type of data contained in the documents to be imaged, and CRT-based units will remain in the forefront at least for the immediate future.

6.2.4 Earth Resources

Of particular interest is the area of earth resources as a concomitant of cartography, where good cartographic data is of prime importance, as well as the data obtained by various imaging means. These data may be obtained and converted into visual form by the use of satellite imaging. The level of resolution required may vary, depending on the data available, and the resulting images can be presented at different levels of resolution. These types of images provide excellent examples of the advantages of both visualization and imaging (VISIM) as used for this application and help to establish some of the requirements for the display. Volume imaging by means

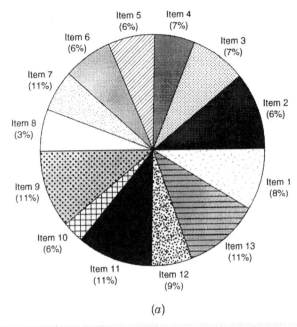

(a)

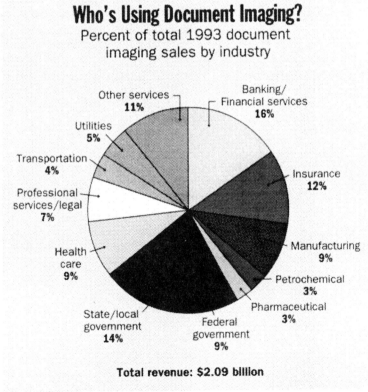

Who's Using Document Imaging?
Percent of total 1993 document imaging sales by industry

Other services **11%**
Utilities **5%**
Transportation **4%**
Professional services/legal **7%**
Health care **9%**
State/local government **14%**
Federal government **9%**
Pharmaceutical **3%**
Petrochemical **3%**
Manufacturing **9%**
Insurance **12%**
Banking/ Financial services **16%**

Total revenue: $2.09 billion

(b)

Figure 6.3 (a) Generalized document imaging data. (b) Specific document imaging data.

of stereo pairs, another capability of present imaging systems, makes it possible to create 3D images of mountains and other volumetric aspects of earth resources. Many other features of earth resources images may be used for agriculture, meteorology, forestry, and firefighting, to name only a few applications. This makes earth resources an important subgroup of the VISIM application for electronic displays. Its equipment requirements are similar to those for cartography.

6.2.5   Geographical Information Systems (GIS)

As noted in Section 5.2.5, cartography is generally the most significant aspect of this application for electronic displays. It is concerned with the production, usually in hard-copy form, of maps and charts, frequently in minute detail, and often including a variety of images to highlight the relevant areas of the map. Thus the requirement is for high-quality plotters for the hard-copy outputs, and electronic display systems with similar capabilities, to allow the desired hard-copy results to be developed and examined prior to their final production. The full range of products listed in Table 6.3, with the possible exception of projectors, may find some use in different aspects of this application. However, the primary initial requirement is for dynamic displays that can adequately show the imagery involved so that it can be determined whether the resulting map or chart will be adequately represented when it is reduced to hard-copy form. The use of projection techniques to further determine the adequacy of the image is a possible option, but it is not essential to obtaining satisfactory results.

6.2.6   Imaging

Although imaging has been combined with visualization to make up a joint application because of the similarities in their display requirements, it is sufficiently important on its own to be treated separately as an application. This is particularly true for complex images such as those found in medical applications, 3D imaging, and a variety of scientific applications. These are covered separately to some extent below, and supplement the brief statements in this section.

6.2.7   Mathematics

Mathematics is an extremely interesting subgroup application for VISIM, in particular the visualization portion. This is because mathematics deals primarily with abstract symbols and numbers. Therefore, the ability to create images that present these abstract elements in a manner that is easier to understand and evaluate than the original can be of significant value. The results of this action can more readily be termed *visualization* than imaging, since neither real-world data nor document imaging is involved. Rather, equations or other mathematical symbology are the sources, and the visualization is the alternate form in which the information contained in these mathematical formulations may be presented. A simple example might be to show the meaning of a parabola by creating its visual form in conjunction with the relevant formula. The same can be done for any other equation that defines some pattern that can be presented visually, such as the double helix that represents DNA. The complexity of the visualization that can be shown on a dynamic display is limited only by the capabilities of the monitor used, and the hard-copy outputs are similarly constrained by the performance limitations of the available systems and equipment for that purpose. The source material is essentially unlimited in its potential complexity when in its original form, so that only the highest level of display performance is adequate to instrument this application group. An interesting example of a high level of mathematical visualization and imaging is the 3D image shown in Figure 6.4. This image represents a geometric solid rendered from a set of 3D equations. Other plots of 2D and 3D formulas can be generated by the appropriate software, and add important contexts to what may originate as abstract concepts.

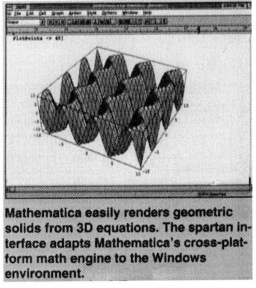

Mathematica easily renders geometric solids from 3D equations. The spartan interface adapts Mathematica's cross-platform math engine to the Windows environment.

Figure 6.4 Image of a geometric solid rendered from a set of 3D equations. Graphic created by Mathematica. Courtesy of Wolfram Research, Inc.

6.2.8   Medical Applications

Medical applications for VISIM are growing rapidly as physicians learn how to use the facilities that produce the visual results. One important group is digital x-ray imaging using the latest developments in AMLCDs, in conjunction with amorphous silicon detectors. Imagers with resolution as high as 1536×1920 and sizes as large as 26×26 cm have been developed to go with development-stage FPDs with diameters as large as 60 cm. Another imaging development that offers high promise is magnetic source imaging (MSI), which can show soft-tissue detail. The system works by using a large number of magnetic field sensors that detect local electrical currents in neural and muscle tissue and is an extremely powerful approach to imaging data that would be difficult to obtain and make available in visual form by any other means.

Another emerging medical application for VISIM, termed *telemedicine*, represents a means for transmitting medical data such as x-rays; and direct medical diagnosis, including surgical simulation by means of a combination of digital imagery and virtual reality. Although these techniques are in their infancy, the promise for remote medical treatment at various levels of complexity is quite large.

6.2.9   VISIM Presentations

Both aspects of the generalized VISIM application have some significance for the presentations subgroup. This is because the activity involved in conducting a presentation includes the use of slides that may be made to visualize a wide range of sources, from numeric data to complex images, as well as imaging the material developed from any of the sources covered in either the subgroups discussed previously, or those presented in the subsequent sections. This makes the presentation application potentially the most encompassing of all the subgroup applications, and, as one result, its requirements cover the range from the simplest to the most complex, depending on what the particular set of slides might include. As a result, it is difficult to arrive at a set of specifications that would be equally applicable to all possible sources. Thus, in some cases a simple slide production system could be adequate, but if all possibilities must be considered, the more complex system—capable of producing high-resolution, wide-color-gamut slides— would be required. The simple system might be no more than a camera that could be used to photograph the image on a monitor that has sufficient resolution and color capability to meet the requirements of the particular presentation, whereas the high-quality slide system must be far more complex. This is discussed in more detail in Section 6.4.1.9 insofar as slides are concerned. However, it is possible to note at this point that the direct-view dynamic display need not be as high-quality in its performance as the slidemaking system as long as it is adequate for viewing the visual material that will be transferred onto slides. This means that lower-resolution FPDs might be used for the initial viewing, although it is probably best to approximate the slide resolution to the furthest extent practical.

6.2.10   Scientific Data

There is a wide range of VISIM applications for scientific data, in particular for the imaging aspect. This application is in some respects similar to the general data or document imaging application, which are quite significant in themselves. However, its importance is further attested to with respect to the visualization aspect by the large number of products and vendors engaged in satisfying the needs of this application subgroup, where software alone accounts for 50 vendors. In addition, a general visualization system (GVS) has been developed at the NASA Ames Research Center for the sole purpose of supporting scientific visualization of data. Further developments of image analysis programs have led to increasing use of displays to both visualize and analyze scientific data. These applications include the use of 3D imaging and motion video for image analysis. However, for this type of application the RS-170 video standard is usually not adequate, and it may be necessary to resort to nonstandard video, although the new high-resolution TV and video systems such as are proposed for HDTV should overcome this deficiency and make video scientific visualization more effective. In any event, there seems to be a productive future for the use of VISIM for scientific applications. It may be some time before non-CRT display systems can meet this requirement, but the main contenders—AMLCD, EL, FET, and plasmas—are expected to achieve this goal.

6.3   TECHNOLOGIES

The technologies for the specification requirements for the VISIM application are essentially the same as those for the computer graphics applications listed in Tables 5.7–5.17. This leaves the technology parameters for the dynamic and hard-copy displays presented in Tables 5.5 and 5.6, to be shown in a somewhat different form in this section, broken down in terms of the "single viewer," "group viewing," and "multiple group" audiences. These user driven forms are shown in Tables 6.4 and 6.5. These distributions of parameter values for the three possible user groups are not necessarily mutually exclusive. However, they do cover the majority of the display applications, and any other groupings should be treated as adjuncts to these main application types. In addition, although the groups are primarily those involved in the visualization aspects, similar distributions of parameter values and application areas also fit the imaging aspects and may be assumed as equally applicable.

TABLE 6.4   User-Driven Dynamic Display Technology Parameters

Technology	Parameter	Single Viewer	Group	Multiple
CRT	Size	15–21	17–31	Not used
	Resolution	640 × 480– 1600 × 1280	800 × 600– 1280 × 1024	
	Color	16–256	250–4000	
	Response (μs)	1–10	5–10	
	Contrast ratio	100	10–100	
AMLCD	Size	9.6–13.1	9.6–13.1	Not used
	Resolution	800 × 600	640 × 480	
	Colors (no.)	256–256, 000	256–256, 000	
	Response (ms)	80	80	
	Contrast ratio	60	20–60	
PMLCD	Size	9.6–11.3	9.6–11.3	Not used
	Resolution	640 × 480	640 × 480	
	Colors (no.)	16–256	16–256	
	Response (ms)	300	300	
	Contrast ratio	10–20	10–20	
FLCD	Size	15	15	Not used
	Resolution	1280 × 1024	640 × 480– 1280 × 1024	
	Colors (no.)	16–256	16–256	
	Response (ms)	10	10	
	Contrast ratio	20–40	20–40	
CRT projector	Size (in.)	Not used	60–240	60–240
	Resolution		800 × 600	800 × 600
	Colors (no.)		16–256	16–256
	Contrast ratio		20–100	20–100
LCD projector	Panel area (in.)	Not used	8.1 × 6.1	8.1 × 6.1
	Resolution		640 × 480	640 × 480
	Colors (no.)		4096	4096

6.4   SPECIFICATIONS AND BLOCK DIAGRAMS

6.4.1   Specifications

6.4.1.1   Introduction

The specifications presented in this section are for the systems and equipment that can be used to meet both the general VISIM applications and all those listed in the application subgroups. To these can be added a large number of other subgroups, not included in Table 6.1, but as these do not add new requirements to those set by the listed groups, the specifications presented in the following sections may be extended by analogy to cover the nonlisted groups. It should also be noted that the technology parameters listed in Tables 6.4 and 6.5 for dynamic displays and hard-copy units, respectively, are equally applicable to both the visualization and imaging

TABLE 6.5   User-Driven Hard-Copy Technology Parameters

Technology	Parameter	Single	Group	Multiple
Pen plotter	Display area (in.)	12 × 12	12 × 12–30 × 40	30 × 40
	Resolution (dot size in.)	0.001–0.005	0.005–0.01	0.01
	Pens (no.)	1–8	1–8	1–8
	Colors (no.)	1–8	1–8	1–8
	Speed (in./s)	1–20	1–4	1–2
	Accuracy (%)	0.01–0.05	0.05–0.1	0.1
Electrostatic	Display width (in.)	24	24–48	48
	Resolution (dpi)	400	200, 400	200
	Colors (no.)	4–16	16	16
	Speed (ips)	0.3–2.5	0.3–2.5	0.3
	Accuracy (%)	0.1–0.2	0.1–0.2	0.1
Inkjet	Resolution (dpi)	180	180–360	Not used
	Speed (cps)	167	167	
	Jets (no.)	12–64	12–64	
	Colors (no.)	256–16 million	256–16 million	
Laser	Resolution (dpi)	240, 300, 400, 600	240, 300, 400	240, 300
	Text speed (ppm)	4–120	4–120	4
	Graphics speed	0.5–5	0.5–5	0.5
	Colors (no.)	16–256	256	256
Thermal	Resolution (dpi)	200–300	200–300	Not used
	Speed (ppm)	1–3	1–3	
	Colors (no.)	16–16 million	16–16 million	
	Size (in.)	8.5 × 11–11 × 17		
Slidemaking	Resolution (TV lines)	800–8000	Not used	Not used
	Colors (no.)	16–16 million		
	Speed	30 s–5 min		

aspects of the total VISIM specifications. They are not repeated here in the interest of conciseness, but may be referred to if necessary. In addition, the technologies–products table shown initially as Table 5.4 has been repeated as Table 6.3 for convenience of reference, as noted in Section 6.2.1. In this case, the display technologies for the VISIM applications are essentially the same as those referred to in Section 5.3.1, so that the repetition of Table 5.4 is appropriate. As noted in that section, they are covered in Chapter 3. These sources may be referred to for more detail on these products, whereas the coverage here is limited to that necessary to understand the requirements and specifications appropriate to the specific applications covered in this chapter.

6.4.1.2   CAD/CAE

These are two of the most widely used applications in the general VISIM group, and have some of the most severe specification requirements. The material given here is restricted to the VISIM aspects, supplemented by the more general treatment in Chapter 9. They are combined here because the requirements are very similar for both, and only small differences in certain

TABLE 6.6   Specifications For VISIM CAD/CAE Applications

Parameter	Value CAD	CAE
a. Dynamic Displays		
Luminance (cd/m^2)	100–300	50–100
Contrast ratio (no.)	10–30	10–20
Resolution (TV lines)	800×600–1200×1024	800×600–1200×1024
Colors (no.)	256–16 million	256–16 million
Viewing area (diagonal in.)	17–27	17–27
b. Hard-Copy Displays		
Resolution (dpi)	200–700	200–400
Speed (ppm)	1–3	1–2
Colors (no.)	4000–16 million	256–4000
Size (in.)	8.5×11–21×40	8.5×11–18×27

parameters need be covered (e.g., cf. parts *a* and *b* of Table 6.6 for dynamic displays and hard-copy units, respectively).

A different range of sizes is shown for CAE than for CAD because the space available for the equipment in the CAE area is usually smaller than for CAD, and more individual units may be provided, so that the larger units can be used for both applications.

6.4.1.3   Documents

"Documents" is definitely one of the most important applications for the imaging part of the VISIM application group, as is attested to by the large number of vendors, where a limited sampling lists over 20 companies; some of the industries that use document imaging are listed in Table 6.7. These percentages are for an earlier year, and surely changed, as did the number of user groups involved. In any event, it is appropriate to focus on the require-ments to ensure that they will meet the requirements of this rapidly growing subgroup of the general VISIM application. To this end, a specification for the dynamic display broken down into the visualization and imaging aspect is given in Table 6.8*a* and for the hard-copy display, in Table 6.8*b*.

It should be noted that the differences between the visualization and imaging aspects of the documents group are somewhat arbitrary, and in general the higher requirements are satisfactory for both. However, this separation is used to indicate that it is possible to apply somewhat lower requirements for imaging if desired. This may lead to a lower cost for the latter, although the loss in performance may be somewhat undesirable. It is up to the user to determine what is adequate for the operation involved, assess the relative costs, and select the lowest performance that can be used effectively.

TABLE 6.7   Document Imaging User Industries

Industry	Market Share (%)
Banking and financial	15
Federal government	12
Health care	6
Insurance	13
Legal and other professional services	5
Manufacturing	20
Petrochemical	5
Pharmaceutical	4
State and local government	10
Transportation	5
Utilities	5

TABLE 6.8   Specifications for VISIM Documents Applications

Parameter	Value	
	Visualization	Imaging
a. Dynamic Displays		
Luminance (cd/m^2)	100–300	50–100
Contrast ratio (no.)	10–30	10–20
Resolution (TV lines)	800×600–1200×1024	800×600–1200×1024
Colors (no.)	256–16 million	256–16 million
Viewing area (diagonal in.)	17–27	17–27
b. Hard-Copy Displays		
Resolution (dpi)	200–700	200–400
Speed (ppm)	1–3	1–2
Colors (no.)	4000–16 million	256–4000
Size (in.)	8.5×11–21×40	8.5×11–18×27

6.4.1.4   *Earth Resources*

As noted previously, the display requirements for this application may be quite high, particularly when high-resolution maps are to be created from the information that is obtained from satellites or other airborne sources. In some cases low resolution is adequate for the purpose as demonstrated by the separation of the requirements data between visualization and imaging, where the first may use lesser capabilities than the second. The two sets of parameter values are given in Table 6.9*a* for the dynamic display and Table 6.9*b* for hard copy. The values for imaging are shown first as this is likely to be the most demanding of the two parts of the application requirements, especially when the end product is expected to be high-resolution maps.

TABLE 6.9   Specifications For VISIM Earth Resources Application

Parameter	Value	
	Imaging	Visualization
a. Dynamic Displays		
Luminance (cd/m^2)	100–300	50–100
Contrast ratio (no.)	10–30	10–20
Resolution (TV lines)	800 × 600–1200 × 1024	800 × 600–1200 × 1024
Colors (no.)	256–16 million	256–16 million
Viewing area (diagonal in.)	17–27	17–27
b. Hard-Copy Display		
Resolution (dpi)	200–700	200–400
Speed (ppm)	1–3	1–2
Colors (no.)	4000–16 million	256–4000
Size (in.)	8.5 × 11–21 × 40	8.5 × 11–18 × 27

However, as mentioned above, the visualization requirements may, in some cases, be as demanding or even more so than the imaging requirements, and the choice for each must be made judiciously.

6.4.1.5   GIS (Cartography)

This application is restricted to the cartography aspect for the purposes of this section, but the limitation should not be significant, as the display requirements for cartography are as severe as for the more general GIS category, which simplifies the discussion. Thus, the specifications for the dynamic display (Table 6.10*a*) and the hard-copy (Table 6.10*b*) units that are appropriate for mapmaking may also be used for the GIS group. These requirements are essentially the same as those given for GIS as part of the graphics arts application group, as the visual aspects for the two groups are quite similar. However, it should be noted that the hard-copy needs are more stringent for the VISIM GIS group, so the upper levels of the range of values will apply in the majority of cases. This is not necessarily the case for the dynamic display, as it is possible to create the maps adequately with lower values of the dynamic display specification, in particular for the number of colors and display size. This may allow the use of somewhat simpler equipment for the initial design functions, and then production of the actual hard-copy maps on the higher quality equipment.

6.4.1.6   Imaging

As noted previously, imaging, including image processing, is a sufficiently important subcategory of VISIM to warrant separate treatment. This application is also growing in popularity and includes a number of categories,

TABLE 6.10   Specifications for VISIM GIS Application

Parameter	Value
a. Dynamic Displays	
Luminance (cd/m^2)	100–300
Contrast ratio (no.)	10–20
Resolution (TV lines)	800–1600
Colors (no.)	4000–16.7 million
Viewing area (diagonal in.)	21–225
b. Hard-Copy Display	
Resolution (dpi)	200–700
Speed (ppm)	1–3
Colors (no.)	4000–16 million
Size (in.)	8.5 × 11 – 21 × 40

ranging from the manipulation of very large amounts of data and the presentation of the results in visual form to the production of visual images that represent diagnostic data of medical conditions and can also be presented in video form for the use of physicians. Part of this range is shown in Table 6.11, but the total is much greater and includes almost any collection of information that is manipulated so that it can be viewed in an accessible form on some type of terminal or large-screen display. This means that image processing becomes an important part of the operation. These are a small fraction of all the possible applications for imaging, and any application where it is useful to transform data into visual images, or manipulate images to allow comparisons and extraction of information, falls into this category. In particular, the ability to apply powerful image processing programs to alter the image and the transmission of the results in compressed form by video channels makes this application group a potentially significant one. However, to achieve satisfactory results, it is necessary to provide a high-performance system, preferably stand alone, although the combination of a server and either X terminals or workstations is an acceptable alternative. Finally, the

TABLE 6.11   Imaging Categories

Data Source	Image	Application
Compressed document	Original document	Document storage
Magnetic scan	Scanned object	Medical
Measurement data	Video	Measurement display
Satellite	Map	Navigation
Still image	Video	Conference
Fingerprints	Processed video	Identification

least demanding case is when both the processing and/or the visual results are minimal and therefore can be satisfied by a relatively simple system. These lead to the requirements shown in Table 6.12.

In general, the full-imaging requirement appears to be most likely to be needed, as listed applications tend to demand high performance, and the other applications that should be considered fall into the same or similar categories.

6.4.1.7 Mathematics

Visualization is the more important part of the full VISIM requirements for this application, because the direct viewing of the dynamic display, with the ability to view the results of converting the abstract symbolic data into a variety of visual forms, is probably the more useful and interesting aspect of this capability. The display specification requirements may vary by a consid-

TABLE 6.12   Imaging Requirements

| Application | Processor Requirements | | Display Requirements | | |
	Speed (MHz)	Memory (Mbytes)	Resolution	Size (in.)	Colors (No.)
Full imaging	66–100	120–540	1280 × 1024	21–27	256–16 million
Partial imaging	30–60	60–120	800 × 600	17–21	256–4096
Simple images	20–25	40–60	640 × 480	15–17	16–256
Full	—	—	High-end workstation		
Partial	—	—	Server/X terminal, workstation		
Minimal	—	—	Server/ASCII terminal		

TABLE 6.13   Specification For VISIM Mathematics Application

Parameter	Value
a. Dynamic Displays	
Luminance (cd/m^2)	100–300
Contrast ratio (no.)	10–20
Resolution (TV lines)	800–1600
Colors (no.)	4000–16.7 million
Viewing area (diagonal in.)	21–225
b. Hard-Copy Display	
Resolution (dpi)	200–700
Speed (ppm)	1–3
Colors (no.)	4000–16 million
Size (in.)	8.5 × 11–21 × 40

TABLE 6.14   Dynamic Display Specification For VISIM Medical Application

Parameter	Value
Luminance (cd/m^2)	100–300
Contrast ratio (no.)	10–20
Resolution (TV lines)	800–1600
Colors (no.)	4000–16.7 million
Viewing area (diagonal in.)	21–225

erable amount, depending on the complexity of the mathematical formulation and the amount of data that may be involved. In addition, the visualization and imaging requirements may differ by a considerable amount, depending on whether hard copy is required and/or visualization is sufficient. In any event, the range of parameter values given in Table 6.13 should cover from the least to the most complex image requirements.

6.4.1.8   Medical Applications

Medical applications are growing rapidly for both dynamic and hard-copy display capabilities. Telemedicine offers considerable opportunities for employing the highest-quality displays, and the imaging capability must meet the advanced requirements of a variety of operations. The values for the full VISIM application are listed in Table 6.14 for the dynamic display.

6.4.1.9   Presentations

Presentations may impose particularly demanding display requirement on the VISIM application, or one that is only nominally difficult. This difference in the requirement depends on whether the slidemaking or group presentation is involved, and in some ways represents the difference between the "single" and "group." Viewing requirements are usually extremely demanding, although there are some slidemaking apparatuses that produce much lower-quality images. However, these use much simpler slide production systems, where the major component is the camera, and do not fall into the display

TABLE 6.15   Dynamic Display Specification For VISIM Presentations Application

Parameter	Value
Luminance (cd/m^2)	100–300
Contrast ratio (no.)	10–20
Resolution (TV lines)	800–2000
Colors (no.)	4000–16.7 million
Viewing area (diagonal in.)	15–27

TABLE 6.16   Hard-Copy Display Specification For VISIM Slidemaking Application

Parameter	Value
Resolution (dpi)	200–2000
Speed (min.)	3–10
Colors (no.)	4000–16 million
Size (mm)	35–70

application category. Therefore, the slidemaking part of the VISIM presentation display requirement is restricted to the high-quality slide production operation, and the presentation portion refers to the group aspect. These are broken down into the visualization for the first, and imaging for the second, as shown in Table 6.15 for the dynamic display and Table 6.16 for the slide portion, representing the hard-copy requirement.

The parameter values for Table 6.16 are for producing the finished slide, and apply to the slidemaking apparatus. It is assumed that no other hard copy is required for the presentation activity; if it is, then it may consist of prints of the slides to the extent considered desirable by the presenter.

6.4.1.10 Scientific Data

This class of data is a somewhat arbitrary choice, and any number of other types of data are subject to the same or very similar display specification requirements. However, scientific data do represent a wide range of data

TABLE 6.17   Specifications For VISIM Scientific Data Application

Parameter	Value	
	Imaging	Visualization
a. Dynamic Displays		
Luminance (cd/m^2)	100–300	50–100
Contrast ratio (no.)	10–30	10–20
Resolution (TV lines)	800 × 600–1200 × 1024	800 × 600–1200 × 1024
Colors (no.)	256–16 million	256–16 million
Viewing area (diagonal in.)	17–27	17–27
b. Hard-Copy Display		
Resolution (dpi)	200–700	200–400
Speed (ppm)	1–3	1–2
Colors (no.)	4000–16 million	256–4000
Size (in.)	8.5 × 11–21 × 40	8.5 × 11–18 × 27

types and may serve for most of the others with respect to these requirements. Thus, this display application group is more than adequate for the data category. In addition, both dynamic and hard-copy displays may be involved, as well as different requirements for the visualization and imaging aspects of the general VISIM requirements (see Table 6.17).

6.4.2 Block Diagrams

6.4.2.1 Introduction
Block diagrams of the equipment and systems used for both applications in the VISIM groups are many and varied (see examples in Figs. 6.5–6.8). Of these, only a few are presented and discussed here as sufficiently representative of all applications to adequately demonstrate the type and complexity of the systems. To further demonstrate the basic system types involved, Figure 6.5 is sufficiently detailed to be used as the basis for discussion. However, a few comments are in order prior to viewing that block diagram. First, in the interests of generality, many of the data sources are combined in only a few categories, although any data source qualifies for the imaging aspect. In addition, visualization applies to any operation that presents some form of image on the electronic display screen. Further, the production of hard copy from either the digitized data or the visualized image is another type of VISIM output that remains important in spite of all attempts to lessen that type of permanent display medium. Finally, all these data and image sources must be subjected to a varying amount of processing, requiring both standard and graphic operations.

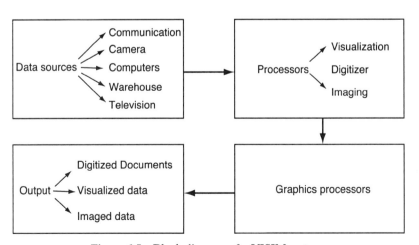

Figure 6.5   Block diagram of a VISIM system.

Thus, the total system may become quite complex, although it is shown in relatively simple form in this generalized system block diagram (Fig. 6.5). In addition, although the output may be shown on only two basic types of equipment—dynamic monitors and printer–plotters—the variety of technologies used for both and the integration of the display with other parts of the system extend the actual system into a multiplicity of specific combinations. This variety of operational equipment grouping is covered to some extent in later sections concerned with more specific configurations used for particular application groups. With this introductory discussion of the meaning of the generalized VISIM system block diagram shown in Figure 6.5, it is feasible to comprehend the meaning and limitations of the blocks shown and proceed next to the specific systems that are used for specific applications, such as document imaging and telemedicine described earlier in this chapter.

6.4.2.2 *Document Imaging*

Document imaging is one of the most widely used operations in the VISIM application groups. The basic form a typical system may take is pictured in Figure 6.6, where the monitor and printer represent the two of the three output systems shown in Figure 6.5. The data source is a document the scanner converts into the digitized image stored in the computer memory.

This system shown in Figure 6.6 may be expanded into larger versions, as shown in Figure 6.7*a*, *b*, that can handle multiple documents in combination with a number of workstations and printers that permit the viewing of the outputs by many individuals at the same time.

Next, the simplified block diagram of a typical PC imaging system is shown in Figure 6.8. The changes between the early 1980s and late 1990s versions

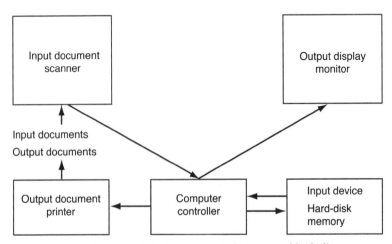

Figure 6.6   Document to image conversion system block diagram.

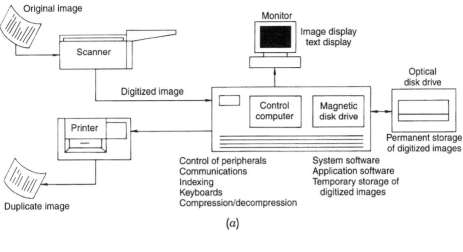

(a)

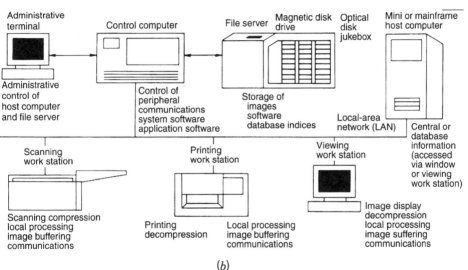

(b)

Figure 6.7   Block diagrams of (a) a typical document imaging system and (b) a large imaging system. Courtesy of *Electronics*.

consist of the addition of a color camera, Pentium computer, and high-resolution monitors of either the CRT or FPD types. These changes highlight the advances in digitizing, image processing, and display that have occurred in this time interval. In addition, numerous specialized imaging systems, such as those for chemical, measurement, medical, satellite, and scientific applications, constitute a group of variations on the basic theme.

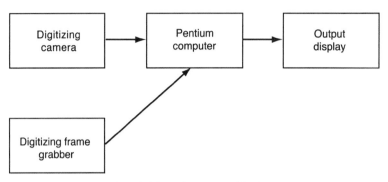

Figure 6.8   PC imaging system block diagram.

Finally, what is probably the most sophisticated form of a document imaging system is represented by the diagram given in Figure 6.9. This diagram represents software from Axxix Corporation termed AXXIS/Bank-File. Here, the multiple document input sources and output displays cover essentially all forms of document imaging and may be considered as the final word at present on imaging systems.

6.4.2.3   Scientific Visualization
Although not the most important example of the visualization application, the scientific version is interesting because of its varied nature, and it is sufficiently diverse to act as a model for other types. The main purpose of the scientific visualization system is to provide a means for presenting data that represents some aspect of scientific investigation in a form that demonstrates meanings that cannot be easily derived from the nonvisualized source.

6.5   CONCLUSIONS

The breadth and range of the applications that fall into the VISIM groups are shown to be very extensive, and the systems available to meet the requirements of these applications are similarly varied. As a concomitant requirement, the electronic display portions of these systems must meet a wide range of requirements, as is demonstrated by the many tables presented in the previous sections of this chapter. From these it may be concluded that all the technologies and equipments that are available can be adapted to one or more of the applications, and advances in the state of the electronic display art will be welcomed by all users.

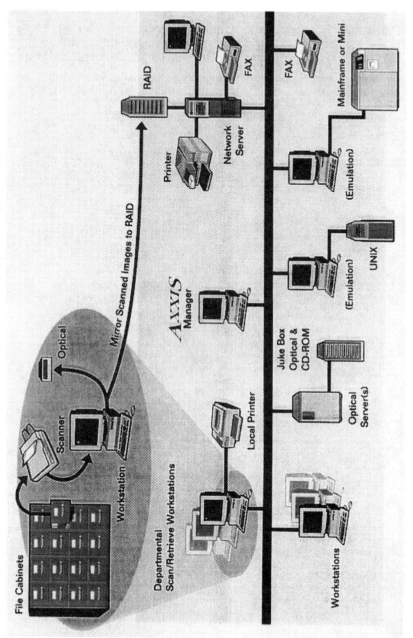

Figure 6.9 Block diagram of high-level document imaging system. Copyright ©1997 by Axxis Corporation.

REFERENCES

1. McCormack, B. H., and Brown, M. D., *Visualization in Scientific Computing*, National Science Foundation, 1987.
2. Machover, C., Ed., *CAD/CAM Handbook*, McGraw-Hill, New York, 1996, Section 5.1, p. 31.
3. Robinson, L., "In Imaging, What a Difference a Decade Makes," *Photonic Spectra* April 1995, 109, 1.

7

MULTIMEDIA AND PRESENTATIONS

7.1 INTRODUCTION

These two applications use one or more types of electronic displays as the primary means for making the outputs of the application activity available for the user. Indeed, the visual outputs are the only manner in which the results of the presentations are made useful, and the same should be said about multimedia, except that there may be other nonvisual outputs as well. Thus, it is expedient to examine these applications in greater detail than most of the others that are covered to a limited extent in previous chapters. To achieve this greater detail, the applications are described and analyzed in the following sections of this chapter in terms of which display requirements best implement the applications and which technologies should supply the preferred type of display device and/or system. The system definition used here is the one given in Section 4.1 and may be applied to any and all product types that are referred to. The technologies and the product types are covered in detail in Chapters 2–4, and are not treated in further detail here, except when certain special features need particular elaboration and discussion. All aspects of the display requirements are included and supported by charts and tables as required for ease of interpretation and comprehension. The discussion is broken into four main parts, beginning with application descriptions, followed by technologies and products, and ending with system specifications and block diagrams.

7.2 APPLICATION DESCRIPTIONS

7.2.1 Multimedia

Multimedia is a term that has come into extensive use for applications that use visual displays, especially with the advent of CD-ROM software, which has allowed a significant extension of what may be achieved by using personal computers. This new capability has allowed the addition of art images, sound, and video directly to the text that might be created. Other capabilities, including some not yet conceived of, may be added to the presently available media types and further enhance what may be done. In addition, applications that do not use displays may also be part of the multimedia installation, characteristic examples of which are voice transmission and music generation. As a result, multimedia is a general capability that can be used effectively by many different specific applications. Multimedia is also considered in some contexts as an application on its own, and this is a tenable position for description and analysis. Actually, it should be defined as a technology that may be used for a multiplicity of applications, some of which do not require the visual display capabilities that are the main subject of this volume. One possible definition is that it is a technique that allows many different applications to use a variety of media, such as animation, graphics, sound, text, and video, to name the most common examples, in many different combinations that allow information to be conveyed in some optimum form. The nondisplay applications may be included without in any way diminishing the role of the display in the other applications that are part of the total multimedia group. Table 7.1 lists a number of these applications, and clearly shows that multimedia is far more than a separate application. This is a formidable list, and each has its own characteristic way of using the multimedia capability. One interesting example of how multimedia is used is in medicine, where surgeons can present a demonstration of surgical procedures by means of programs that allows a range of these procedures to be shown on a monitor screen and include animation, stills, and interaction allowing a user to move through the different anatomic layers. These images may be used for presentations, demonstrations, teaching, and patient information. Video may also be included, and printouts of any image may be

TABLE 7.1   Multimedia Applications

Application	Displays	Other Multimedia Units
Animation	Monitor, plotter	Audio, computer
Computer	Monitor, plotter, printer	Audio, CD-ROM
Data transmission	Plotter, printer	Receiver, computer
Education	Monitor, projector	Audio, CD-ROM, video
Editing	Monitor, printer, plotter	Computer, CD-ROM, video
Kiosks	Monitor	Computer, touchscreen
Medicine	Monitor, projector	Audio, CD-ROM
Presentations	Monitor, projector	Audio, CD-ROM, video
Publishing	Monitor, plotter, printer	Computer
Television	Monitor, projector	Audio, receiver, VCR
Video	Monitor, projector	Audio, CD-ROM, VCR

produced. Another less esoteric example of a successful application of multimedia, is the use of kiosks in supermarkets that allow the customer to obtain information about a variety of aspects of the available products. The kiosks consist of monitor displays with touchscreens that allow the customer to ask questions and obtain answers. The multimedia aspect of the system is the access to extensive databases that maintain information about products and facilities and the interactive aspect that gives the customer the ability to create displays that satisfy the need for specific information. Indeed, the availability of a multiplicity of databases for a variety of applications is one of the most pervasive aspects of multimedia use. These databases are supplemented by a host of authoring programs that help to create the sequence of operations needed for the specific application. These programs run on Windows and Mac platforms, and the results can be used on the same platforms. Typically, audio, animation, video, and text can all be part of the resulting program, which can be tailored for the specific application.

This description of some of the applications for which multimedia may be used provides a short introduction to the value of the multimedia capability when used in a careful and fully professional way. This result is best achieved by allowing an expert in the field to author the multimedia sequences. However, it should also be noted that the hardware and software must also be of the best quality and provide adequate performance capabilities for satisfactory results to be obtained. These capabilities are discussed later.

7.2.2 Presentations

Presentations make up another broad category, similar to multimedia in that there are a number of separate applications. Although presentations may be considered as an application on its own, they may also be used as or aid or supplement to other applications. However, beginning with presentations as an independent application, it consists of—as the term implies—presenting a series of visual images that are usually accompanied by an oral text. It is generally used to describe a sequence of slides that are used to inform, educate, entertain, or influence one or more viewers, but may also refer to a video display, and could include cinematrographic techniques, primarily movies. Thus, "presentations" is a term that applies to a variety of formats and is a versatile application for displays, electronic or otherwise, as well as an aid or supplement to other applications, as noted. This situation further highlights the similarity to multimedia, and this is enhanced by following a similar procedure in this description, namely, listing the most important applications that may use presentations as an adjunct, as is done in Table 7.2. This is a relatively brief list, but does include most of the major applications that may use presentations as part of the general operations. However, it should be recognized that presentations are basically for the purpose of communicating to groups of individuals, and any application that requires the

TABLE 7.2   Presentations Applications

Application	Displays	Other Presentation Units
Briefings	Monitor, projector, plotter–printer	Audio, CD-ROM, computer, VCR, video
Education	Projector, printer	Audio, CD-ROM, VCR, video
Entertainment	Projector, plotter–printer	Audio, CD-ROM, VCR, video
Government	Monitor, projector, plotter	Audio, VCR, video
Marketing	Projector, plotter–printer	Audio, computer, VCR
Military	Monitor, projector, printer	Audio, computer, video
Sales	Projector, plotter–printer	Audio, computer, VCR, video

ability to perform this communication may benefit from having some type of presentation capability. The display type is primarily some form of projector, but monitors may also be practical when the number involved is not too large to be awkward or overly expensive, although individual monitors have been found feasible in some airline installations. Another example of the use of separate displays is for the presentation of translations of opera librettos, although this application generally uses projectors. It should also be noted again, in concluding this discussion, that presentations have many of the same characteristics as multimedia, and the most successful presentations include most of the same capabilities and equipment as those found in multimedia installations, so that the two applications have much in common and can be viewed as essentially synonymous in terms of displays.

7.3   DISPLAY TECHNOLOGIES

7.3.1   Introduction

The discussions and descriptions contained in the following sections supplement those found in Chapters 2 and 3, and are limited to those capabilities and characteristics that impact most on the two general applications that are the subject of this chapter. Therefore, only limited detail is included, with reference to the expanded discussion in the previous chapters when required. It is important to note that these two applications have rather stringent display requirements, and the technologies and products that may be used are representative of the best performance available. It should also be recognized that when other factors such as cost and availability are considered, it may be necessary to accept lower performance capabilities, and this is recognized in determining to what extent less-than-optimum performance may be adequate. It should be understood that the best performance should be the goal.

The display technologies and products found in both multimedia and presentations applications are quite similar, and may be discussed together

TABLE 7.3   Technologies versus Products

Product	Technology	Advantages
Flat-panel	AMLCD, PMLCD	Color, low power
	EL, FED, plasma	Luminance, size, cost
Monitor	CRT, FPD	Data density, color, size, cost
Plotter–printer	Pen, electrostatic	Size, quality
	Inkjet, thermal	Quality, color
Projector	CRT, LCD	Size

for the purposes of this analysis. In addition, the two topics may be combined for the purposes of this section, as the technologies are directly related to the products in so far as any specific technology is used for a specific type of product. Thus, there is no loss of generality in this approach, and the particular aspects of either topic may be given adequate coverage with this approach. As an introductory step, it is useful to list the technologies in conjunction with the products that use them, and this is done in Table 7.3. This list is restricted to the specific display technologies and products used for multimedia and presentations and does not include any of the other products that may be used for total systems, such as are listed in Table 7.2. They are covered in some detail in Chapter 3, and are treated again in Section 7.4 in relation to their performance capabilities as they impact on the operational requirements for multimedia and presentations applications.

7.3.2   Hard-Copy Technologies

As noted in Section 5.3.3, the hard-copy technologies for these applications are used to produce prints and other images for use when further study is required. These images range in complexity from simple to quite elaborate, and slides or film may also be used. Therefore, these hard-copy technologies are included in Table 7.5, which is essentially the same as Table 5.6, repeated here for convenience of reference. The technologies are quite similar for both application groups and are combined for the purposes of analysis of the parameters.

7.4   PRODUCTS

7.4.1   Dynamic Displays

Most, but not all of the technologies covered in Chapter 2, and listed in Table 2.1, are involved in the electronic displays used for multimedia and presentations applications. The most popular one is the color CRT, particularly in its implementation for CRT monitors. This has also been the case for

TABLE 7.4   Dynamic Display Technology Parameters

Technology	Parameter	Multimedia	Presentations
CRT	Luminance (cd/m^2)	30–300	30–300
	Display size (diagonal in.)	17–31	21–56
	Resolution (H × V)	640 × 480–1280 × 1024	800 × 600–1280 × 1024
	Colors (no.)	256–16 million	256–16 million
	Response time (μs)	1–10	1–10
	Contrast ratio	20–100	50–100
AMLCD	Display size (diagonal in.)	9.6–13.2	9.6–13.2
	Resolution (H × V)	640 × 480	640 × 480
	Colors (no.)	256–4096	256
	Response time (ms)	10–100	50–100
	Contrast ratio	10–30	10–50
PMLCD	Display size (diagonal in.)	9.6–13.2	9.6
	Resolution (H × V)	640 × 480–800 × 640	640 × 480
	Colors (no.)	256–4096	256
	Response time (ms)	100–300	300
	Contrast ratio	10–20	10
EL	Luminance (cd/m^2)	10–50	10–60
	Display size (diagonal in.)	13–24	13
	Resolution (H × V)	512 × 256–1024 × 864	640 × 480–1024 × 864
	Color (no.)	16–256	1–16
	Response time (ms)	1–10	1–10
	Contrast ratio	10–20	10–20
CRT projector	Luminance (cd/m^2)	60–300	60–300
	Display size (diagonal in.)	60–240	60–100
	Resolution (H × V)	1280 × 1024	800 × 600–1280 × 1024
	Colors (no.)	256–16 million	256–16 million
	Contrast ratio	20–100	10–100
LCD projection	Panel area (in.)	7.9 × 5.3	8.1 × 6.1
	Resolution (H × V)	720 × 480	640 × 480
	Colors (no.)	256–4096	256–4096
Plasma panel	Luminance (cd/m^2)	20–50	20–100
	Panel diameter (in.)	10–40	20–60
	Resolution (H × V)	640 × 480–1280 × 1024	640 × 480–1280 × 1024
	Colors (no.)	256–260,000	256–260,000

projectors, but LCDs have begun to take over the projection function. Similarly, for FPDs the AMLCD is in the forefront for those applications, although PMLCD does show some promise. Table 7.4 contains a list of the most important parameters and the range of performance that is required for either multimedia or presentations applications. Much of this information may be found in Table 5.5, but it is recast in Table 7.4 in terms of the requirements for the multimedia and presentations application.

TABLE 7.5   Hard-Copy Technology Parameters

Technology	Parameter	Multimedia and Presentations
Pen Plotter	Display area (in.)	$12 \times 12 - 30 \times 40$
	Resolution (dot size in.)	0.001–0.005
	Pens (no.)	1–8
	Colors (no.)	1–8
	Speed (in./s)	1–20
	Accuracy (% of movement)	0.01–0.05
Electrostatic	Display width (in.)	24–48
	Resolution (dpi)	200, 400
	Colors (no.)	4–16
	Speed (ips)	0.3–2.5
	Accuracy (% of line)	0.1–0.2
Ink Jet	Resolution (dpi)	180–360
	Speed (cps)	167
	Jets (no.)	12–64
	Colors	256–16 million
Laser	Resolution (dpi)	240, 300, 400, 600
	Text speed (ppm)	4–120
	Graphics speed (gppm)	0.5–5
	Colors (no.)	16–256
Thermal Transfer	Resolution (dpi)	200–300
	Speed (ppm)	1–3
	Colors (no.)	16–4096
	Size (in.)	$8.5 \times 11 - 11 \times 17$
Slide Making	Resolution (TV lines)	800–2000
	Colors (no.)	16–4096
	Speed	30 s–5 min.

7.5   SPECIFICATIONS

7.5.1   Introduction

The specifications described and analyzed here are for systems that are adequate for both the multimedia and presentations group of applications. They may be used to define the performance requirements for both and as a data source for the preparations of operational requirements for the hardware and software. In particular, the performance capabilities of the dynamic and hard-copy displays are of prime interest at this point. These are covered in a general way in Tables 7.4 and 7.5, and are developed for specific examples in the application groups for both multimedia and presentations as shown in Tables 7.1 and 7.2. Several examples are selected from those shown in these tables, and representative specifications presented for the displays used in these applications, with appropriate discussions of how the specifications are chosen from the possible performance range.

TABLE 7.6   Dynamic Display Specification For Multimedia and Presentations Applications

Parameter	Value
Luminance (cd/m^2)	30–300
Contrast ratio (no.)	10–50
Resolution (TV lines)	600–1600
Speed (in./s)	1–5
Colors (no.)	256–16.7 million
Viewing area (diagonal in.)	17–27
Memory size (Mbytes)	40–120

Next, to supplement the information contained in the specifications, block diagrams of the systems involved in producing the output displays are given so that the general form of the implementation may be reviewed and described. These descriptions include all the most significant characteristics of the hardware, software, and operational procedures involved in producing and using the systems needed to meet the operational requirements of the specific applications. As an introduction to the individual multimedia applications, the general display requirement is presented first at this point in Table 7.6.

7.5.2   Multimedia Applications

7.5.2.1   Education

Education has become an important multimedia application with the advent of a variety of computer programs that can be used as an aid in a classroom with a teacher, or as a means for students to obtain information about a wide variety of subjects, and carry out independent learning activities using personal computers and CD-ROMs. The latter have been most effective in teaching mathematics to children aged 4–12 and 8–12 years, and use monitors with at least a 13-in. diagonal display and 256 colors. The programs allow the children to create 2D and 3D objects in various forms, as well as rotating them and viewing them from a variety of angles. Other programs use techniques developed for videogames to make learning attractive, and keep children interested for hours. The teaching potential of these programs is very high, and can be extended to the high-school and university levels. This approach supports the emphasis that is being paid by educators and government officials to providing computer systems in all schools, and connecting these systems to the Internet. The impact of these programs on future educational techniques is potentially enormous, and should lead to a widespread use of electronic displays for this application.

Although the specifications shown in Table 7.6 should be more than adequate for the display requirements of the education application, cost

TABLE 7.7   Dynamic Display Specification For Education Application

Parameter	Value
Luminance (cd/m^2)	10–100
Contrast ratio (no.)	10–20
Resolution (TV lines)	480–800
Speed (in./s)	3–5
Colors (no.)	16–256
Viewing area (diagonal, in.)	13–21
Memory size (Mbytes)	40–120

considerations will probably lead to a lesser performance capability being adequate. This is shown in Table 7.7.

7.5.2.2   Medicine

Medical applications are important examples of how electronic displays may be used in teaching, training, and carrying out the activities involved in a large number of medical procedures. One example is an interactive program termed "animated dissection of anatomy for medicine" (ADAM), which enables physicians to demonstrate surgical procedures to patients on the display screen. It is interactive, and versatile enough to be used as a teaching tool for either high-school or medical students. The procedure is to show a life-size human male on the screen by means of a highly accurate illustration, and then allow the viewer to move interactively through the layers of the anatomy down to the bones. Animations of surgical procedures may also be presented and viewed on the screen. The program has been designed by medical illustrators, and the visual material is capable of being quite graphic. In addition, in a slightly different form, the program is designed for home use as an educational tool. Therefore, in order to achieve maximum benefit from the program, it is desirable to have the best possible output displays. The program can be operated on both Macintosh and PC-compatible computers with CD-ROM drives. The very high quality of the visual images imposes an extreme requirement on the displays, as shown in Table 7.8.

TABLE 7.8   Dynamic Display Specification For Medical Application

Parameter	Value
Luminance (cd/m^2)	30–300
Contrast ratio (no.)	10–50
Resolution (TV lines)	1000–1600
Speed (in./s)	1–5
Colors (no.)	256–16.7 million
Viewing area (diagonal in.)	17–27

The hard-copy units, when required, should have similar resolution values, but the sizes may be much larger, depending on which technology is used. The available types are listed in Table 7.5, and any of the technologies may be used, with the pen and electrostatic plotters best for large-size inkjet and thermal transfer for number of colors, and slides for multiple viewing.

7.5.2.3   Kiosks

Kiosks signify a facility that permits the usage to access a variety of information sources at one or several locations; they are particularly useful in providing location data in department stores and supermarkets, among other retail operations. In addition, they may be used in hotels to direct guests to various facilities and shops in the hotel, with the latter paying part of the purchase costs. The equipment consists of a terminal with a touchscreen input that enables the user to select from an on-screen menu a variety of source and location data. The information may be stored in a CD-ROM or other digital data memory, and display on either a CRT or FPD monitor. Video, photos, animation, audio, and art may also be included, depending on the exact type of service desired. This is an extremely versatile example of the use of multimedia in commercial applications, and may be extended to education in museum locations.

Because of the wide range of specific applications for these kiosks, it is difficult to establish an exact set of display specifications for the units. However, it is possible to define a general type of usage that suffices for the majority of uses, assuming that the direct retail application is sufficiently representative for this purpose. In this case, the requirement for luminance is fairly high, but resolution may be limited so that AMLCDs with high background illumination may be acceptable. Such a specification is shown in Table 7.9, where the matrix data applies primarily to FPD versions and the speed is the time needed to display a full screen. As stated previously, these are average requirements that should be adequate for the majority of applications, and they may be extended for certain special applications for which higher parameter values are necessary, in particular for resolution,

TABLE 7.9   Dynamic Display Specification For Kiosk Application

Parameter	Value
Luminance (cd/m^2)	300
Contrast ratio (no.)	50
Resolution (TV lines)	1000
Speed (s)	5
Colors (no.)	256
Viewing area (diagonal in.)	17
Matrix $(W \times H)$	640×480

number of colors, and matrix size. These changes may be readily introduced to make the specification adequate for a wider range of applications without changing the basic structure of the specification. One example of an application needing such change is a museum in which visitors view the artwork on a kiosk before going to the actual location of the art.

7.5.2.4 Training

Training, especially for corporate purposes, has become an important example of the use of multimedia to aid in carrying out the teaching operation and introducing employees to a variety of concepts and activities of significance to the organization. These corporate training operations are significantly more expensive than simple presentations used for briefings, but they can be much less costly than instructors, particularly if the use of the latter requires travel to some of the job site, such as training centers. In addition, multimedia courses may be prepared by outside consultants, including interactive capabilities, and then used effectively throughout the organization. The electronic display portion of the multimedia equipment may be used for a variety of multimedia activities, such as videotape and CD-ROM. The requirements for the display are also quite varied, ranging from the minimum used with a desktop computer, to the extensive capabilities associated with high-resolution projection systems. This range is shown in Table 7.10.

Hard-copy material often is part of the training operation and may be in a variety of formats and produced by using many of the available hard-copy devices. In addition, a full publication activity may be involved, resulting in a text that is provided to the participants. This leads to another use for multimedia—that is, publication—but this application is not discussed further here as it does not add information significantly different from that presented previously. In any event, the training and other applications described are sufficient to illustrate the importance of multimedia, and illustrate what some of the specification requirements might be for the dynamic display.

TABLE 7.10   Dynamic Display Specification For Training Application

Parameter	Value
Luminance (cd/m^2)	100–300
Contrast ratio (no.)	30–50
Resolution (TV lines)	640–1600
Speed (s)	1–5
Colors (no.)	256–16 million
Viewing area (diagonal in.)	12–256
Matrix $(W \times H)$	640×480–1600×1200

7.5.3   Presentations Applications

7.5.3.1   Introduction

Presentations as a display application was, until recently, the primary means for preparation and display of slides and transparencies as illustrations for lectures, briefings, and similar activities that are enhanced by the use of visual material. For these applications, the major capabilities required are those for the preparation of slides, transparencies, and other visual material. These are satisfied by the various slide production systems classified as standalone workstations with slide projectors added for the actual presentation. In addition, the use of dynamic display projectors has enabled the user to access images created digitally and stored in memory. The latter adds computer graphics to the system, so that it could be classified as similar to that described previously. Therefore, for the purposes of this description, the slidemaking system alone is considered as primarily a standalone workstation and the multiple projection system as a client–server/X Windows configuration. In addition, the advent of LCD projection systems has added the ability to create images on line for presentations as well as limited slides and stored images. The display and system requirements, along with possible systems are shown in Table 7.11.

The presentations group of applications for electronic displays is much less varied than the multimedia group, but has many features in common with the latter. Indeed, to some extent presentations employ multimedia techniques and might be considered as one example of multimedia. These considerations noted, it is still convenient to consider it as a separate group. To some extent, slidemaking may be considered as part of the presentations group, with the quality of the slides as one of the important effects on the display requirements. However, in cases where the slides use a standard projector, the specifications are those of that unit and relate mostly to the optics, whereas when a video projector is involved the characteristics of the display on the projection unit, whether it is a CRT or an LCD panel, also

TABLE 7.11   Presentation Requirements

Application	Processor Requirements		Display Requirements			System Requirements
	Speed (MHz)	Memory (Mbytes)	Resolution	Size (in.)	Colors (No.)	
Multiple projection	66–100	120–540	1280 × 1024	21–27	256–16 million	Server/X terminal
Slidemaking	30–66	60–120	1280 × 1024	17–21	256–4096	High–end workstation
LCD projection	20–60	40–60	640 × 480	15–17	16–256	Server–workstation

enter into the requirement. It is the latter case that is of most interest and is treated in more detail here. Other types of presentations used for the education application are those that employ computers and VCRs, which are somewhat similar to slide projectors that use LCD panels that are controlled by computers. These exhibit display requirements that are quite similar to those for the slide projectors and may be combined with them in a single table as shown in Table 7.11.

7.5.3.2 *Education*

Education is one representative presentation application. In its simplest form education may be limited to only one source of presentation data, that is slides, and one type of display, namely, projectors. This example of a presentations application most frequently uses the simplest type of projection display in order to minimize cost, although for large lecture classes a more elaborate presentation means may be employed, such as some of the CRT units covered in Section 3.5, or an LCD projector with better capabilities. A complete list of projector types may be found in Table 3.31, and the ones listed in that table should be adequate for most uses.

Various types of video projectors use a CRT or a light valve as the projection image source, and several of these are described in Section 3.5. Of these, the ones that use an LCD panel as the light valve have become the most popular for the simple presentation application covered here. Table 7.12 is an appropriate specification for this example of a presentations application.

7.5.3.3 *Briefings*

Briefings constitute one of the most widely used examples of a presentations application. In its actual specific form, it ranges from the simplest review or lecture to elaborate perorations on a broad group of significant topics. These topics may include important information on topical events or information on industrial matters pertinent to the organization or group for which the briefing has been prepared. In its most elaborate form, these briefings take on many of the characteristics of a multimedia activity, and the presentations

TABLE 7.12 Dynamic Display Specification For LCD Projection System

Parameter	Active Matrix	Passive Matrix
Colors (no.)	256–4096	16–256
Resolution (pixels $W \times H$)	640×480–1024×1024	640×480–800×640
Contrast ratio	10–50	10–20
Video input	NTSC/PAL/SECAM	NTSC
Size (diagonal in.)	8.4–13.4	8.4–9.6
LCD panels (no.)	1–3	1

group becomes part of the multimedia group. This may lead to some confusion in classification, as noted previously, but is retained here to illustrate some of the differences between the two groups.

The simplest form of presentation is the lecture that uses only a fully optical slide projector with no video capability, and a group of slides that have been prepared in advance and may be advanced at the appropriate time by either the lecturer or the projectionist if union rules prevail. The only interactive aspect is when new visual material is created on transparencies by the presenter, and this may be either in the course of the lecture or in answer to a question. This procedure is a far cry from multimedia, and is increasingly rare as computer-driven dynamic displays become more generally available as presentation equipment. Therefore, this relatively simple approach is being replaced, even when only a straightforward presentation is involved, by the computer-driven projection systems using CRTs or light valves as the visual source. In these systems, the optical components may be similar to those used in the optics-only projectors, but the addition of the computer system with appropriate programming and input devices creates a far more flexible vehicle, and makes even the simplest presentation far more effective, even without the multimedia capability. The equipment required for this level of presentation is covered in the next section, but it may be noted here that it consists primarily of some type of personal computer in conjunction, in most cases, with a standard optical projector that uses a liquid crystal panel as the slide portion of the operating system. This system may then be used in the same fashion as an optics-only system, with the slide material previously developed and stored in the computer memory, and called up to the LCD panel as required. The final display parameters may be at minimal levels, depending on the capabilities of the LCD panel. It is also possible to attain better performance by using a CRT or light-valve projector so that high levels of visual capabilities are available, even for this type of relatively limited presentation system. This is illustrated in Table 7.13. This type of application assumes either a large room or lecture hall with space for up to 100 participants. Larger audiences could require some higher levels of luminance and/or contrast ratio, but this might mean going to the best-performing

TABLE 7.13 Dynamic Display Specification For Briefings Application

Parameter	Value
Luminance (cd/m^2)	50–100
Contrast ratio (no.)	10–20
Resolution (TV lines)	640–1200
Colors (no.)	16–256
Viewing area (diagonal ft)	10–25
Matrix ($W \times H$)	640×480–1200×1024

light-valve systems, and is more appropriate for the more elaborate corporate briefing or high-level sales presentation, which usually require more total capability, and begin to approach a true multimedia application. This application represents the next level of presentations and, as noted, could be considered a full multimedia example. However, for the purposes of this section, it is treated as an application that falls into the lower category represented by the presentations application group. This approach allows a convenient separation between presentations and multimedia application groups.

7.6   BLOCK DIAGRAMS

The block diagrams presented and discussed in the following sections are representative of what is available in terms of equipment and systems to meet the display requirements generated by the multimedia and presentations groups of applications. Due to the similarity of requirements for the two groups, it is possible to cover them both in the same section, with similar block diagrams, although the multimedia requirements and systems go beyond those generated by presentations. The block diagrams shown include all of the equipment that may be found in any one of the systems used for either multimedia or presentations applications, each labelled as to the application involved.

The display system includes a variety of inputs that are connected to an RF distribution system from which some of the inputs are received. Beginning with an example of such a distribution system, a block diagram is shown in Figure 7.1.

The next step in designing a complete multimedia system is to introduce other sources of data beyond those available from the system shown in Figure 7.1. These are depicted in Figure 7.2, and include a number of different sources not directly connected with the terminal, but capable of being fed into it by means of various modifications.

The additional inputs to the terminal are provided from such sources as a camera, and a scanner, databases from some mass storage media, all combined through high-speed I/O ports, as well as an image processor and a Pentium computer processor. This broad array of sources may be added to those shown in Figure 7.1. To these may also be added those immediately connected to the terminal such as a VCR and CD-ROM. From Figures 7.1 and 7.2, it is clear that the multimedia sources are many and varied, and lead to a highly flexible form of data acquisition and display. To these many capabilities could be added film, other video, and voice to complete the totality of possible sources.

Finally, for an example of one popular multimedia application, there is the kiosk type, used in many supermarkets and department stores as information

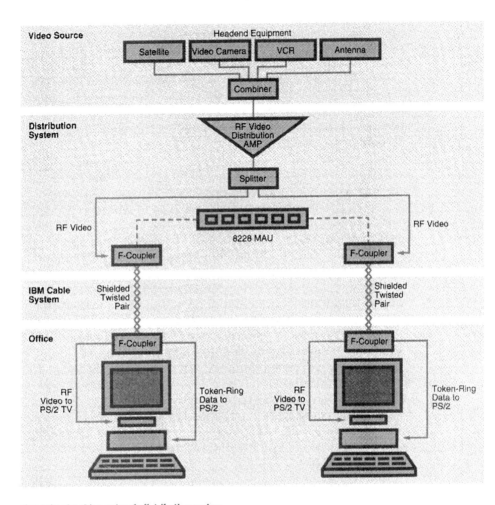

Example of a video network distribution system.

Figure 7.1   Block diagram showing typical multimedia RF distribution system. Courtesy of International Business Machines, Inc.

sources for the customers. This is shown in Figure 7.3, and is representative of what may be found in many locations.

Each kiosk may be used for different purposes, such as the store directory, or the produce available at particular locations. In addition, they may contain other relevant, or sometimes irrelevant, information, depending on the whim of the store manager, including sales and other offerings. The information is contained in the basic database located in the mainframe or other computer

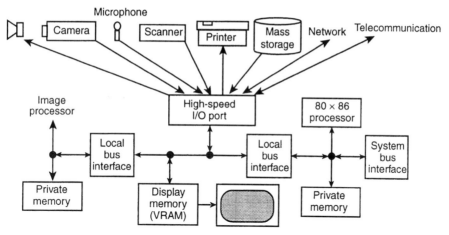

Figure 7.2   Block diagram showing types of multimedia inputs. Courtesy of *Electronic Design* (Jan. 7, 1993).

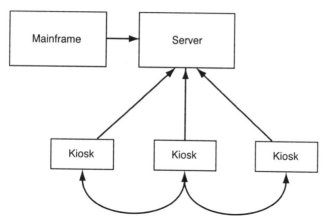

Figure 7.3   Example of a multimedia application using a kiosk.

types capable of storing large amounts of data, and the server contains the software required to operate the whole system as an information source.

Although, as noted previously, systems for presentations-type applications can reach the complexity associated with multimedia, they are frequently considerably less complex, containing only a few of the many data sources associated with full multimedia systems. One such presentations system, namely, the one used for straightforward lectures or paper presentations, is shown in Figure 7.4, and constitutes what is probably the simplest form of such a system, if the one with only an optical projector is ignored.

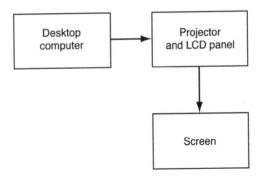

Figure 7.4   Block diagram of the simplest type of presentation system.

Figure 7.4 is a deceptively simple block diagram as the combination of the desktop or laptop computer and the LCD projection panel allows some rather sophisticated actions to be performed, such as modifying the display images and adding new images. However, it is still a relatively simple system in that the data sources are quite limited, although the lecturer can introduce new material or respond to questions by altering existing material.

Finally, there is the full presentations system that includes several other sources, in particular VCRs, CD-ROMs, and other video material. This type of system might qualify as an example of a multimedia system, and is in some ways the crossover point between multimedia applications and presentations applications systems. As noted previously, it is possible to combine the two groups, but for the purposes of this chapter it is convenient to separate them, which to some extent creates a distinction with only limited difference. However, at least in its simplest form the presentations applications system does differ recognizably from even the simplest multimedia system, and therefore it is worth treating them separately. This final system is shown in Figure 7.5.

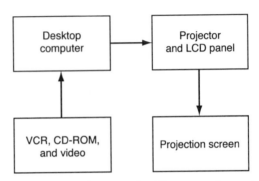

Figure 7.5   Block diagram of a full presentation system.

Figure 7.5 appears to be a simple block diagram, but its complexities are hidden by the inclusion of three important data sources. If these were expanded, the block diagram would be far more involved. However, for the purposes of this discussion, it is sufficient to show them as controlled by the computer, which then outputs the combined images directly to the projection panel. It is also possible to feed their outputs through a combiner directly into the projection panel, but this is somewhat more limiting insofar as system flexibility is concerned. In any event, this example of a presentations system completes the discussion and block diagrams of the systems involved.

8

ENTERTAINMENT: ELECTRONIC GAMES AND VIDEO

8.1 INTRODUCTION

Electronic games (EG) and video (CD-ROM, TV, VCR, laser disk) are combined in this chapter under the general rubric of "entertainment," although it might be argued that the first creates too much stress in some of its forms to qualify, and that the latter at least purports to offer information as well as entertainment, although the former frequently contains much of the latter. Regardless of these somewhat questionable divisions, it is convenient to combine these two applications as their display requirements tend to be quite similar. In addition, the common elements of the application, as indicated by the general descriptive term of entertainment, make it useful to treat them both in a similar fashion. Another important reason for combining the two applications, is that they both tend to use the same display technology and system configurations, and therefore the technology has a significant impact on the system display requirements. This similarity simplifies the task of reviewing and describing the technologies involved, as well as the manner in which the operating systems are defined. The procedure in this chapter is similar to that in the previous chapters on user applications, beginning with application descriptions. However, it is not necessary for these descriptions to be preceded by a list of specific applications beyond the two that make up the subtitle of this chapter, as these are sufficiently inclusive in themselves to cover all applications relevant to the general entertainment category. Of course, the subcategories of CD-ROM, TV, and VCR do warrant some separate treatment as their electronic display requirements are somewhat different, although the basic technology is quite similar insofar as the characteristics of the information signal are concerned. Therefore, they are broken out in the subsequent sections whenever there are specific differences in the requirements that are restricted to particular subcategories. Thus, VCR and laser disk may differ in some parameter such as resolution, and both may differ from CD-ROM. However, the trend is to have the same player for a variety of CD-ROM and laser disk players, so that they may reasonably be combined for most sets of requirements. This is done as indicated by the requirements and the need for simplifying the text.

8.2 APPLICATION DESCRIPTIONS

8.2.1 Electronic Games

8.2.1.1 Introduction

Electronic games may be divided into three basic categories that may be defined as (1) handheld, (2) PC-operated, and (3) arcade-based, all of which use some kind of computer to control and operate the actual game. This differentiates them from the mechanical types, and necessitates some type of output electronic display to make the game feasible.

The specification requirements for the display may differ for the three different types of games, sometimes by noticeable amounts. However, the best performance is certainly adequate for all and can be used if desired, although in certain cases, in particular for the handheld games, the demands of low cost and small size may predominate. This may change as palmtop units become less expensive and take over more of the game market. However, the very low-cost games will surely continue to be served by units designed specifically for single games, and here the cost advantage will be too great to allow more generalized systems to compete. Thus, it may be anticipated the three types will continue to exist for some time, and examination of the performance requirements for all of them is warranted. However, to understand the reasons for the differences, it is necessary to describe some of the electronic, mechanical, operational, and other physical characteristics of the three types. This is done in the following sections.

8.2.1.2 Handheld Games

The general appearance of a handheld electronic game is of a small plastic case containing the electronics to operate the game controller that activates the plug-in game cartridge. It is battery-operated and has an LCD screen with limited resolution and color capabilities. The main sources for this type of game are Nintendo and Sega, and an extensive number of cartridges can be played through the basic controller. The small size and minimal LCD screen allow for only a limited number of colors and pixels. Thus, even the best designs are subject to visual limitations, as well as operational difficulties. The display specifications are given later in Table 8.2, and at this point it is sufficient to note that the screen size is in the order of 2–4-in. diagonals and the number of colors, between 16 and 64. This is admittedly inadequate for satisfactory viewing, but the advantages of portability and small size seem to outweigh the severe limitations on performance, at least for children, who are the main consumers of this type of product. The alternative is to use the computer-driven versions of the handheld games that provide all of their characteristics at much better performance quality.

8.2.1.3   Computer-Driven Games

As noted previously, the handheld games are also available in computer-driven form, where the cartridge is replaced by floppy disks or CD-ROMs. In addition, a large number of other games are available in the same format, and can be played on a variety of PCs, ranging from the 386s through the Pentiums, and including the other IBM types and Apple Macintosh. In these cases, the performance capabilities are the same as can be achieved for any of the other electronic display applications. These games are designed to be played on some form of PC, and in general are available on either floppies or CD-ROMs, although the trend is toward the use of CD-ROMs, especially where the amount of hard-disk memory required is very large, which is increasingly the case for popular games such as Myst. However, the use of CD-ROMs requires the availability of a physical driver, which adds to the complexity of the PC unit. The usual procedure is that a certain portion of the CD-ROM program is transferred to the hard disk for permanent storage, and the rest is kept on the CD-ROM for transfer when the game is being played. This allows for a smaller permanent storage requirement, which is becoming increasingly important as programs designed for use with Windows 95 require larger storage capacity. This frequently leads to the requirement of as much as a gigabyte of hard-disk memory, which fortunately is not too expensive, especially if obtained when the equipment is first purchased. Similarly, CD-ROM drivers capable of as much as 25 × drives may be obtained at reasonable cost and provided excellent animation. Therefore, it may be anticipated that the CD-ROM games will take over the market. Further evidence of this trend is seen in the temporary availability of much lower-cost versions of CD-ROM games on disks, and the disappearance of floppies.

Given this expected trend, one might ask what the effect will be on the electronic display requirements. It can be concluded that the parameter values will be much more stringent, as higher resolutions and numbers of colors can be readily handled by CD-ROMs. These values can be met by the majority of available PCs, and further demonstrate the advantages of computer-driven games, at least insofar as satisfactory video images are concerned. In addition, the proliferation of palmtop and similar units may permit the use of portable units for this application, although the costs will greatly exceed those of the simpler handheld games.

8.2.1.4   Arcade Games

Electronic games found their first significant acceptance in arcades that contained a number of different types in association with large screens, and relatively inexpensive game playing costs. The most popular types were the war games, sports, and car racing, where the quality of the visual presentation was such as to create a sensation of reality. These games had much in common with the imagery produced by simulation systems, and tended to use the technology developed for such systems, albeit in somewhat less complex form. The visual quality afforded by examples of simulation and game displays is quite impressive and calls for ultra-high-performance electronic display equipment and systems. In addition, although modern PCs have sufficient performance capabilities to meet the demands for high-speed response, high resolution, and multicolor capability in the operating system, there is still place for other types of computers, in particular servers, for these specialized game applications. The performance capabilities of arcade games are quite extensive and varied, leading to the high parameter values given later in Table 8.4.

However, with all the proliferation of computer-driven arcade games, the most exciting aspect is the introduction of virtual reality (VR) techniques into the generation of electronic games. VR is treated in some detail in Chapter 10, and further information on its application to games may be found in Section 10.2.7. The capabilities afforded by the application of VR techniques to electronic games are quite extensive, ranging from the use of helmet-mounted displays (HMDs), covered in detail in Section 10.3.2.2, that create realistic illusions of complicated activities in conjunction, with both PC and mainframe computers, to the simplest simulation of much less complex games activities, such as 3D chess. For these activities, entertainment HMD display parameters, listed later as Table 8.6, and in Chapter 10 as Table 10.5 demonstrate one type of HMD that can be used for electronic games. These parameters are somewhat arbitrary, in that the design of HMDs is still in a state of flux, but they are representative of what has been achieved at present. In addition, there are other approaches to achieving the type of 3D display needed for true VR, such as the ones that use special glasses, as well as those that achieve the 3D effect by different optical means such as switching the image between two versions that differ by the amount needed to achieve the desired effect. The parameters for these are of the best quality.

8.2.1.5 Summary

It is apparent from the previous discussions that there may be large differences in performance capabilities among the three basic types. These are usually determined by the types of users involved, as well as other considerations more concerned with business aspects. Thus, the handheld games are generally confined to young users, and are purchased as gifts by older relatives and friends. The computer games are customarily associated with computer users, who are frequently adults who use the computer for other purposes, with the games as a pleasant diversion. Indeed, the games sometimes intrude on more serious applications, if it is reasonable to believe certain anecdotal comments. In addition, juveniles are beginning to use computers for more serious purposes such as homework and references, so the same conflict may exist for them. Finally, the arcade games are strictly for entertainment, and are frequently largely by teenagers. In any event, the range of available performance capabilities is wide, and it is convenient to show this comparison by combining the specifications as is done later, in Table 8.5.

8.2.2   Video

8.2.2.1   Introduction

One definition that has been proposed for video information found in television systems is that it is "data conveying a description of a picture which can be displayed by an appropriate reproducing system through the use of television signals" [1]. Another definition, again limited to television, is found in the *IEEE Standard Dictionary*, and states that video is a term pertaining to the bandwidth and spectrum position of the signal resulting from television scanning. In order to expand the term somewhat from those limiting it to television, "video" is used here as a general term for all manifestations of the visual presentation of electronic signals that may consist of a variety of forms of data and information. This presentation will result from the conversion of the electrical signal into visual form by various means, and the signal itself may be either analog or digital, at least in its original form.

The sources of the video signal are numerous and various in the technologies used to create the electrical signals that correspond to the original and final versions of the image. The most prominent example is the TV signal that is transmitted to a receiver and then converted into the video signal. The transmission may be directly from the station by wireless, indirectly from a satellite, or directly by cable. In each case the signal that reaches the receiver is essentially identical insofar as the video portion is concerned, and it is the function of the receiver to make the necessary conversions so that the imaging device can be properly actuated. Once converted into the proper video format, whether NTSC, PAL, SECAM, or RGB, the video signal may be applied to the imaging device, and the path from the source is completed. Details on the techniques for coding and decoding the color video signal using any of these techniques may be found in Benson's handbook [1, Chapter 4], and for the purposes of this discussion it suffices to note that the video signals applied to the display system can be properly converted into the appropriate form from any of the signals referred to previously. The other sources that should be considered when discussing the video application of electronic displays are the computer memory, TV camera tubes, the laser disk, CD-ROM, and the ubiquitous VCR. Here the signals are stored in solid state, disk, or magnetic media, and are transmitted to the display system by means of light or magnetic pickups where they appear in the form of video signals that correspond to the data contained on the storage medium. Table 8.1 contains the list of sources and the types of storage used for each.

TABLE 8.1   Video Sources

Source	Storage Medium	Bandwidth (MHz)
Computer	Hard disk	10–100
Television	Transmitter	4–10
TV camera	Solid state	10
Laser disk	Hard disk	10
VCR	Magnetic memory	4–6
CD-ROM	Hard disk	10–100
Video disk	Hard disk	100–500

Other sources may evolve in the future, but these are the main ones at present, and are sufficient for the discussion of video applications for electronic displays. Most significant among the electronic display requirements are the ability to accept the 100-MHz bandwidth signals that may be generated by the computer, and which have become a component of the TV signal as HDTV reaches its final state, now that the FTC (Federal Trade Commission) has accepted the digital system developed by the consortium. This may lead to much more demanding parameter values for both the TV-transmitted and TV camera sources. In addition, this advance in TV performance capability will be reflected in improved performance requirements for the laser disk, VCR, and any other medium that stores the image in a form that is converted to video, and then transmitted to the display system. The parameter values are similar to those listed later in Table 8.4 for the arcade games application, and may exceed those as the HDTV performance increases and begins to approach that achieved by photographic presentations such as slides or cinema, which may be anticipated in the not-so-distant future.

8.3   SPECIFICATIONS AND BLOCK DIAGRAMS

8.3.1   Specifications

8.3.1.1   Introduction
The parameter values contained in the specifications for the electronic games and video applications have been referred to previously, and are presented in the following section to establish what the performance requirements are for the display systems and equipment used for these applications. The specifications are presented in terms of a number of significant parameters that are applicable to the performance requirements of the applications, and follow the same procedure as is used for the specifications shown in previous chapters, with the same provisos as to the meaning of the particular parameters and what impact they may have on the total display performance. Thus, they should be considered as representative of what is desirable, but should

not be treated as denying the possibilities that better performance may become available, or that lesser performance may be adequate for certain special situations. For example, in the event that HDTV becomes prevalent and leads to improved TV equipment, the video requirements should tend toward the upper limit, and the same situation will apply to laser disks, and CD-ROMs. In addition, the performance of all types of electronic games may be subject to the same improvements. In any event, the specifications shown in the next section for both electronic games and video may be considered as representative of what can be achieved and used at least as a starting point for establishing display requirements for both application groups that fall within the entertainment application category.

8.3.1.2 Electronic Games

Electronic games exhibit a wide range of performance requirements, beginning with the handheld types, and extending to the arcade games, including virtual reality types. As a starting point, the performance capabilities of the handheld types are shown in Table 8.2. These games tend toward the lower end of the parameter values shown in Table 8.2, in order to keep the costs low, and because the user groups consist mainly of children and young adolescents who are not very demanding in their performance requirements. However, this may change as the performance capabilities of the display technologies used improve in quality and are available at lower cost. However, the range shown will probably remain for some time.

Next in performance capabilities come the PC-operated games, which are frequently the same as the handheld games except for the larger screens and improved performance available on the PC monitors or display screens used in conjunction with the various types of PC units, ranging from the desktop to the palmtop models. The parameter values for these are given in Table 8.3. The improvements in performance are largely in the color range, resolution, and screen size leading to the higher parameter values given in Table 8.3.

Last, but by no means least, there are the arcade games that provide the highest level of performance available, usually more that can be obtained with standard PCs, but the gap is rapidly decreasing as the PCs improve in both operational and display performance, with sufficiently reduced cost so that the higher-performance capabilities may be attained at reasonable price

TABLE 8.2   Dynamic Display Specification for Handheld Electronic Games Application

Parameter	Values
Colors (no.)	16–64
Resolution (pixels $W \times H$)	300×250
Contrast ratio	3–5
Size (diagonal in.)	2–4

TABLE 8.3 Dynamic Display Specification for PC-Operated Games Application

Parameter	Value
Colors (no.)	256–16 million
Resolution (pixels $W \times H$)	800 × 640
Contrast ratio	10–20
Video input	NTSC, RGB
Size (diagonal in.)	9.6–32

TABLE 8.4 Dynamic Display Specification for Arcade Games Application

Parameter	Value
Colors (no.)	4000–16 million
Resolution (pixels $W \times H$)	1280 × 1024
Contrast ratio	20–30
Video input	NTSC, RGB
Size (diagonal in.)	21–50

levels. In any event, the specifications for arcade games are available now, and are shown in Table 8.4. Finally, it is convenient to show the comparison among the handheld, computer-driven, and arcade games, by combining the specifications in Table 8.5, which repeats the information given in the three previous specifications and makes them available for easy examination. These parameter values are essentially the same as those shown in Tables 8.2–8.4, and are repeated here in a single table to facilitate comparison of the values for each of the three electronic game types. It is clear from this comparison that the range of performance parameter values is quite large, and may be expected to increase even further as the performance capabilities improve, and the best performance may be achieved even with the standard PC types, so that arcade performance will be possible at home or in the office.

This section on display specifications is completed by presenting those for the VR type that uses some form of HMD. These are shown in Table 8.6,

TABLE 8.5 Dynamic Display Specification for Electronic Games Application

Parameter	Handheld	PC-Operated	Arcade-Based
Colors (no.)	16–64	256–16 million	4000–16 million
Resolution (pixels $W \times H$)	300 × 250	800 × 640	1280 × 1024
Contrast ratio	5–10	10–20	20–30
Video input	NA	NTSC, RGB	NTSC, RGB
Size (diagonal in.)	2–4	9.6–32	21–50

TABLE 8.6   HMD Display Parameters

Parameter	Value
Field of view (FOV)	84°
Display type	TFT LCD
Matrix (V × H)	720 × 240
Dot pitch (V × H, mm)	0.365 × 0.158
Color pixel format	RGB delta
Response time (ms)	80
Contrast ratio	30
Display dots	172, 800
Video signal	Composite/analog RGB

and may represent the electronic games of the future. They may now be found in arcades, and may well be available for individuals once the prices drop for HMDs and other displays that offer stereo capability.

8.3.1.3   Video

The video application covers a number of sources, but the range of performance parameter values is applicable to all sources, so it is reasonable to combine the specification requirements in one table, as is shown in Table 8.7. This table should cover the different types of video signal as well, including that developed for HDTV, which is probably the most demanding. Other video sources, as they become available, may extend the parameter values to some extent, but not enough to make that shown in Table 8.7 invalid or inadequate to establish the performance capabilities that can be obtained from available electronic display equipment and systems.

These parameter values should be adequate for all the expected applications of the video group within the general entertainment application, even including the more elaborate VR games and other activities that will add to the enjoyment of using this advanced technology. Further information on the manifold VR applications may be found in Chapter 10.

TABLE 8.7   Dynamic Display Specification for Video Application

Parameter	Value
Colors (no.)	4000–16 million
Resolution (pixels $W \times H$)	1280 × 1024
Contrast ratio	20–30
Video input	NTSC, RGB
Size (diagonal in.)	21–50

8.3.2   Block Diagrams

8.3.2.1   *Introduction*

The equipment and system block diagrams that apply to the entertainment applications in general, and the games and video subgroups in particular, are shown in Figure 8.1. It may be noted that the output displays are varied, containing all the electronic display types used with TV sets and computers, as well as the more specialized ones such as HMDs, which are used for the VR applications. In addition, several of the input sources may bypass the computer on their way to the screen. This creates a more versatile system, and one that is oriented toward the entertainment applications. Of prime importance is the inclusion of means for converting and transmitting TV images, from the transmitter through the receiver, and more directly from several of the other sources such as the TV camera, VCR, laser disk, and the digital video disk (DVD), also known as the digital versatile disk. This new digital video disc will probably supplant many of the other sources, even including those used for the electronic games, but it is still somewhat too new to make more than hopefully well-informed predictions as to this future application. In any event, the multiplicity of sources for games and video necessitate that the system represented by the block diagram contain all of the elements necessary to operate within the entertainment application environment, leading to the greater complication shown in Figure 8.2.

It should be recognized that Figure 8.2 is a generalized diagram, with no attempt to show all the specific elements involved. However, it is adequate to indicate what the basic data flow is, and what is necessary to go from the data sources to the displays. This diagram may be elaborated to a considerable degree if it is desired to include all the operational elements involved, and this is done to a limited extent in the following sections. This diagram may not include all the possible video types, but the major ones are there, and any

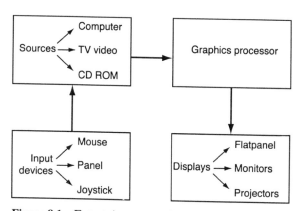

Figure 8.1   Entertainment applications block diagram.

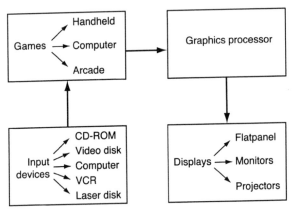

Figure 8.2 Generalized entertainment system block diagram.

additional ones that become available should not add significantly to the elements of the diagram in terms of their impact on the electronic display specifications. Therefore, this generalized block diagram of the elements required for the entertainment specification should be sufficient to indicate what the equipment and system components need be for this application group.

8.3.2.2 Games

Handheld The block diagrams for the three basic types of games are somewhat different for each one, primarily because the sources differ and the data processing capabilities are those appropriate to the particular user group and display capabilities. The diagram for the handheld games is shown first, in Figure 8.3. It is a relatively simple diagram, as the circuitry for this game type is rather restricted, and is designed to work only with the plug-in cards that are used to define the specific game procedure. The types of games that fall into this category are much simpler than the ones designed to run from computer programs, and it is the combination of the instructions contained on the cartridge that acts as the data input source, whereas the actual game assembly performs the necessary data conversion and display functions called for by the cartridge. As the game assembly is designed to operate from a number of cartridges, the functions contained on the card must be somewhat limited and generalized.

At present the only practical display technology is the basic passive-matrix color LCD flat-panel because of both the low power requirements and the simplicity of the display structure. It is to be anticipated that active-matrix units will prevail in the near future as their power requirements decrease so that battery operation for a reasonable period of time will be possible. In addition, laptop units will probably be used for playing handheld games with

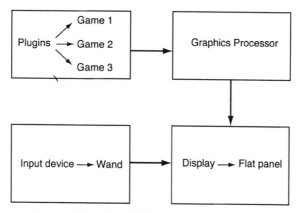

Figure 8.3   Handheld games block diagram.

a full complement of programs, as soon as their costs are reduced sufficiently to compete in the market. Finally, special-purpose computers designed for game playing will continue to be the prime operating means for playing handheld games of more than minimal complexity.

Computer-Driven   Computer-driven games are probably the most successful of the possible approaches to electronic games, especially if the special-purpose computer types mentioned previously are included. The approach that uses a standard desktop computer in association with a multiplicity of programs is the most versatile, especially when the variety of game types available is considered. The type of computer that may be used is quite open, ranging from the lowest-performance desktop to the most elaborate Pentium-based unit. The requirements for the computer performance for any specific game are determined by the game software, and may be quite simple or very complex. In any event, computer-driven games are available for a wide variety of computers, and offer a similarly wide variety of game types, ranging from the simplest ones suitable for young children to the most complex that may be played by mature adults with a great deal of playing skill. Of course, the most successful performers are the adolescents who make up the main market, many of whom are of the "hacker" type. Regardless of the somewhat negative image of these individuals, they have done much to make such games successful. The basic block diagram for the system used to play these computer-driven games is shown in Figure 8.4.

The FPD may use any of the standard technologies, although the LCD is still favored for the PC electronic display. The monitor may use either CRT or FPD technologies, although the CRT is still usually preferred when the display is not an integral part of the computer. However, this situation is rapidly changing as more monitors that use other FPD technologies become available at competitive prices.

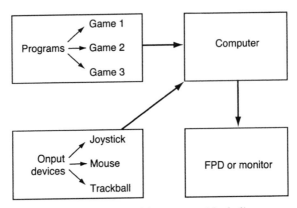

Figure 8.4   Computer-driven games block diagram.

Arcade-Based Systems   Arcade-based systems are to a considerable extent synonymous with the PC types of computer-driven systems. The differences lie in the more elaborate computers that may be employed that are capable of driving more than on installation, and in the higher-performance displays, which can differ in both size and resolution from those used with the single computer-driven systems. This situation leads to the block diagram for the arcade systems to be very similar in many cases to the types illustrated previously. However, in the case of the more elaborate systems that may be found in arcades, there are some possible differences that are significant enough to warrant showing a somewhat different block diagram. In this case, the diagram may include some types of client–server installations, a variety of display types at each station, and space for multiple players at a single station. All these differences lead to a variety of user types to be involved, with some tendency toward the involvement of older individuals, even including adults, mature or otherwise. The block diagram for such a system, shown in Figure 8.5, contains the major components of these systems.

Of course, Figure 8.5 is a simplified block diagram of the arcade system, in that it may include some quite complex and sophisticated games, with correspondingly advanced input devices and electronic displays. In addition, with the advent of VR as a feasible vehicle for games, the possibilities at arcades are significantly enhanced in that the costs of the equipment required, and the availability of space makes the arcade approach most advantageous. A number of sources for VR game installations have been already provided and located at a number of sites. More detail about VR technology and its applications to entertainment may be found in Sections 10.2.6 and 10.2.7, devoted to VR and its applications.

8.3.2.3   *Video*

Video is a broad application area, with multiple input sources, as is shown in Table 8.1. As a result, the block diagram must include a number of configura-

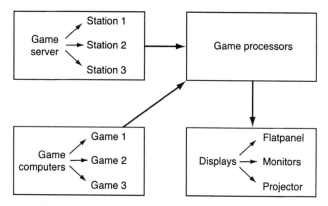

Figure 8.5   Arcade games system block diagram.

tions that take into account all of these sources. This might lead to unacceptable complications in the diagrams, but fortunately all of these sources can be handled by approximately the same combinations of equipment so that the block diagrams can be simplified, albeit somewhat at the cost of completeness. With this admittedly somewhat limited approach in mind, the block diagram shown in Figure 8.6 may be considered as essentially representing all that is generally significant in equipment required to configure and drive an adequate electronic display system for video applications. Changes in the system components are occurring as more and improved sources and output devices become available, but the major functions to be implemented remain the same, and are reasonably adequately represented in Figure 8.6.

The block diagram in Figure 8.6 is subject to considerable change as new digital video sources and players become available, but the same basic

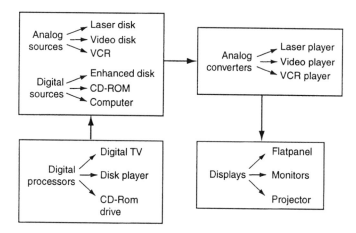

Figure 8.6   Video applications system block diagram.

configuration should be retained, with the data flow continuing as shown. In particular, the enhanced digital video disks and players should have a significant impact on the display requirements as resolution and color capabilities are advanced. The net results of these improvements in both equipment performance and data source content should be performance enhancement to at least and perhaps better than photographic film capabilities. Digital TV is one very important data source that will undoubtedly contribute significantly to the video electronic display requirements.

8.4   CONCLUSIONS

The entertainment application has broad implications for electronic display requirements, and contributes significantly to the performance specifications that must be met to achieve adequate results in the total-presentation systems. The increased capabilities to be found in the various forms of FPDs, monitors, and projectors should lead to greatly improved performance of the display portions of the systems, as well as the multiplicity of data sources that are involved. This becomes of particular importance as such specialized activities as those involved in the many complex games become available, and will surely impact the somewhat less demanding video presentations. Therefore, although the entertainment application, and its subgroups of games and video, including TV in the latter, are not the most demanding of all the display applications covered in this text, it is surely far from the least, and may be used as something of a standard for all the others.

REFERENCE

1. Benson, K. B., Ed., *Television Engineering Handbook*, McGraw Hill, New York, 1986, Section 4.1.1, p. 4.3.

9

COMPUTER-AIDED DESIGN, ENGINEERING, AND MANUFACTURING (CAD/CAE/CAM)

9.1 INTRODUCTION

Computer-aided design (CAD), computer-aided engineering (CAE), and computer-aided manufacturing (CAM) are combined here because their electronic display requirements are very similar and as general applications they are usually lumped together. They make up a triumvirate that impose a wide range of display requirements, and are at least among the most demanding of the various applications that use displays to a significant extent. Computer-integrated manufacturing (CIM) is frequently added to these three, but this application is more concerned with the computer than with the display, does not add anything significant to the display requirements and thus, is not included here with CAD, CAE, and CAM. However, it should be noted that CIM requires less electronic display capabilities than do any of the other three, and therefore this exclusion should not be expected to change the display requirements to any significant extent for this application group.

CAD/CAM is a common way to combine two of the most important computer-aided activities. However, for the purposes of this chapter, it is more reasonable to use the CAD/CAE grouping, as design and engineering are much closer in their electronic display requirements than are the other two. Therefore, the discussions that follow will begin with CAD, proceed through CAE, and conclude with CAM with CIM implied to the extent that its display requirements duplicate those of CAM. This approach allows all the relevant aspects of this group of display applications to be covered in a reasonably complete and meaningful way. It should be realized that each of them has experienced extensive changes and developments from their early beginnings, and the performance of the systems and equipment used for any one or all four has been improved to a remarkable extent. A very important part of this improvement, in particular for the CAD/CAE group, has been in the display aspects, where the great increases in picture quality resulting from high resolution and multiple color capabilities have allowed the outputs to take on appearances that accurately show the characteristics of the final design. This capability is being further advanced by the application of virtual reality (VR) to the design and engineering activities, as described to some extent in Chapter 10, supplements the more limited discussion of the technical aspects of VR introduced in the present chapter, and is intended only to illustrate how VR enhances and improves the CAD/CAE operations. To a considerable extent, as a result of these advances in technology and operations, it is possible to go from concept to final product design and manufacturing in a fraction of the time previously necessary. For example, in the automotive industry it is feasible to produce new models in a year or less, instead of 5–10 years, the same average period, as well as multiple models with only small changes in both the product design and the manufacturing process. This capability also exists for many other types of products, and has enabled vendors to offer the cornucopia of products that represent the market in the advanced industrial countries. It is also to be anticipated that this situation will continue to improve, and make the same bonanza available to the less fortunate in the foreseeable future, with somewhat unforeseen consequences.

9.2   APPLICATION DESCRIPTIONS

9.2.1   Introduction

Although there are similarities in the characteristics of the three applications described in the following sections, there are a sufficient number of differences that affect the electronic display requirements to warrant separate treatment of each, at least at present. Of course, as the cost savings involved decrease, if lesser performance is accepted, these differences may not be sufficient to warrant establishing separate sets of parameter values for each. However, it is still useful to indicate the minimum performance that is acceptable for each application type to ensure that there is no tendency to either over- or underspecify any single one of the CAD/CAE/CAM trio of closely related applications.

   An interesting approach to defining the range of functions that fall within the CAD/CAM classifications is that introduced by the well-known computer graphics consultant, Carl Machover, and outlined in Figure 9.1, where the different actual operations that fall into the application groups are shown under the specific applications, and related to the other operations and information sources that are needed to carry out the functions. It should be noted that although CAE is not shown separately in this diagram except for

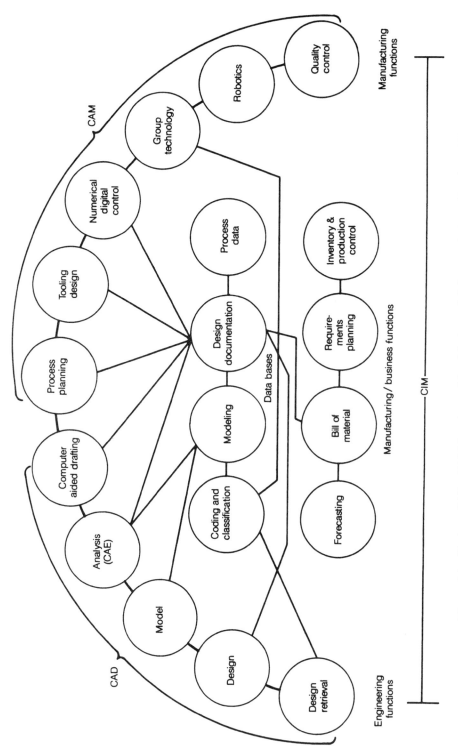

Figure 9.1 Various CAD/CAM functions. After Machover [1], by permission of McGraw-Hill.

291

the analysis function, its functions are very close to those presented for CAD; the major differences are in the form of the outputs that result from the two activities. These differences arise from CAD being largely involved in producing design documents, and CAE confined primarily to mathematical and logical operations that ensure that the design satisfies the operational requirements as much as possible. These specific operations are described in greater detail in the following sections, which are specifically concerned with describing each application in turn.

9.2.2   Computer-Aided Design and Engineering (CAD/CAE)

Computer-aided design and engineering, most frequently referred to by their acronyms, CAD/CAE, are the first, and perhaps the most significant, of the three computer-aided operations that have become such important aspects of product design and manufacture. The earliest use of computers to aid in both electrical and mechanical design probably goes back to the early 1960s, when the results of mathematical calculations and tables of empirical measurements were combined by means of computer operations into output tables that could then be produced in printed form for further analysis and use. The role of the computer was to conduct the mathematical operations and combine the results as needed with other data, thus providing relatively concise listings, in printed form, of the results of the design analysis. This was helpful, especially when complex calculations were necessary, and the hand calculator was not adequate for the job. However, the lists provided nothing more than the results of the calculations, and the computer was used primarily as a high-level calculator. Some attempts were made to present the data in graphical form, but the results were quite limited, and hardly justified the time and expense involved. In addition, it was necessary to wait until the computer and printer had completed their tasks before the simple graphics could be examined. This was all most exasperating, as any one who has gone through this process can testify.

It was not until the advent of the graphical workstation, with either self-contained intelligence, or the ability to connect to some type of mainframe computer, that direct control of the mode of data presentation became available and the advantages of this mode of CAD/CAE became apparent. The advantages of the graphical workstation are indicated to a limited degree by the block diagram shown in Figure 9.2, which is a somewhat simplified version of Figure 4.3, differing only in that it introduces the input and hard-copy devices into the diagram. It is introduced at this point in order to illustrate these advantages. They consist primarily in making it possible to manipulate images and symbols in a convenient manner so that the characteristics of the design can be ascertained in a visually meaningful way.

The hard-copy devices may not be an integral part of the workstation proper, but they are essential components of the total workstation complex and are therefore included in the block diagram. Similarly, the input devices,

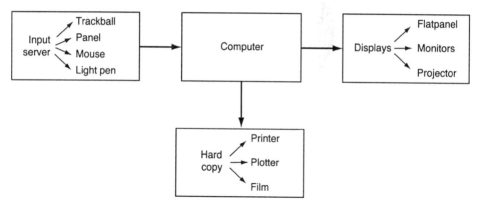

Figure 9.2 Block diagram of graphics workstation.

although attached in some way to the computer, are separable units. However, without some form of input device as an integral part of the total facility, the graphics workstation is essentially inoperable. More detail on the output devices may be found in Section 3.4, and the input devices are covered in Section 4.3.3. One popular CAD application is in the design of electrical and electronic circuits and wiring diagrams. The circuit configuration may be shown on the screen in the form illustrated in Figure 9.3. This view allows the configuration to be examined and analyzed prior to reduction to an actual layout, so that no unnecessary effort is expended on creating the manufactured product. Prior to the advent of CAD, this and similar activities were carried out by means of a long, difficult, and tedious sequence of design activities, without adequate assurance that the design would be accurate, and requiring successive fabrications until the quality assurance was completed. The savings achieved in time and expense, and the improved quality of the product due to the use of CAD, have been important factors in motivating the extensive use of CAD in essentially all design and engineering activities, especially those leading to product manufacturing, with this final operation carried out by the various CAM operations that have evolved, and are covered in the next section.

Another other important feature of the graphics workstation is the multiplicity of input devices that are shown in Figure 9.2, at least one of which—and possibly all—may be included in the total workstation configuration. As noted in Section 4.3.3, the four shown in Figure 9.2 are the most popular, but several others have found such acceptance, and are listed in Table 4.7. In addition, some discussion of the technologies used and the characteristics of the various input devices may be found in that section. However, with respect to the role of the input devices in the CAD/CAE application pair, it may be said, without fear of contradiction, that the workstation would be useless without at least one, and preferably several, of the input devices listed. The reason for this requirement is that an important

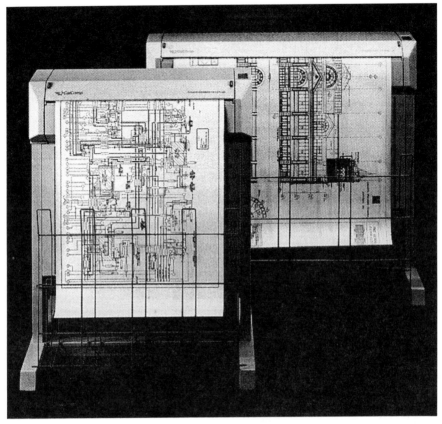

Figure 9.3   Example of circuit configuration diagram. Courtesy of CalComp.

part of the CAD/CAE operation is to allow the workstation to control the location and modify the appearance of the image on the screen of the display system, whether it is CRT- or FPD-based. These considerations and how they impact on the specification requirements of the display system are covered to a limited extent in Section 6.4.1.2, and the material presented here supplements that found in the earlier discussion. In addition, the specifications for the dynamic display and hard-copy devices as they relate to the VISIM aspects of the CAD/CAE applications may be found in Table 6.6. A more detailed specification for a graphic workstations is shown in Table 9.1. New and improved units are constantly being made available, so these specifications are subject to change. However, they are sufficiently representative to indicate the minimum level of performance required for each application, and any better performance can only improve the results. The differences between the CAD and CAE requirements are due to the lower cost that can be achieved with the less stringent values, and may be upgraded to the CAD

TABLE 9.1   Graphics Workstation Specification

Parameter	Value	
	CAD	CAE
Processor (MHz)	100–133	66–100
RAM (Mbytes)	16–256	4–16
Hard disk (Mbytes)	540–5000	540–1000
Display adapter	SVGA, TIGA	VGA, super-VGA
CD-ROM (Mbytes)	Quad, 8 × 660	Quad 660
Input devices	Keyboard, mouse, panel	Keyboard, mouse
Monitor	Table 6.6*a*	Table 6.6*a*
Hard copy	Table 6.6*b*	Table 6.6*b*

values if the cost differences are minimized, and the CAE performance requirements are increased because of a more extensive CAE operation. This latter may occur if the CAE operation is expanded beyond the basic analysis and system engineering functions, and begins to take on some of the actual end-design activities. This trend may lead to more combinations of the CAD/CAE applications into a single operation.

It is of some interest to examine examples of the displays found in CAD applications in order to assess the potential complexity of the images that lead to the electronic display parameter requirements for both the dynamic and hard-copy display systems. Many of these design activities are related to mechanical objects, and Figure 9.4 shows a representative group of design drawings that have been produced on a hard-copy unit after being initially designed using some type of workstation or other type of computer display system. It can be seen that the parameter requirements may be quite demanding to meet the image complexity, especially if a color capability is desired. These display parameters are essentially the same as those shown in Table 6.6 for the VISIM (visualization/imaging), and are repeated with some elaboration later in Section 9.3 to demonstrate the full set of parameter requirements for both CAD and CAE. The reasons for the difference are illustrated to some extent by Figure 9.4, which presents a group of representative images that might be generated for CAE, specifically in those examples for scientific and engineering analysis. The various graphs and mathematical data visualizations are generally less complex than many of the mechanical CAD images, and as a result a computer display system with lesser parameter requirements than for the CAD applications may be adequate for many CAE applications. It should be noted that in some cases the requirements for CAE may approach those for CAD so that equivalent capabilities may be needed. In general it is safe to assume that the lesser capability is adequate, and the use of a higher-performance system may be limited to special cases such as highly complex schematics and wiring diagrams.

Figure 9.4 Example of mechanical design drawings. Courtesy of CalComp.

9.2.3 Computer-Aided Manufacturing (CAM)

CAM is the logical follower to the CAD/CAE sequence in that it uses the output of this pair to carry out the required manufacturing operations. These do not necessarily require a full CAM system, but the rapid development of automatic, computer-controlled manufacturing systems has led to widespread use of CAM facilities for a host of manufacturing operations. One of the earliest examples of CAD/CAM types of operation is the numerical control (NC) machines that are driven by NC programming to automatically produce the part defined by the program. There are several different NC programming systems, and the system is used to prepare a tape that can drive the NC machine. This programming operation is one bridge from CAD to CAM, and has achieved widespread acceptance because of the reduced cost and higher quality that result from the use of this type of automation. Thus, NC programming has gone beyond the programming of machine tools, with which it began, and has expanded to include a variety of machining, turning,

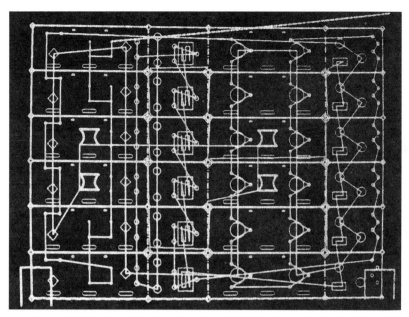

Figure 9.5 Example of an automated nesting program. After Machover [2], by permission of McGraw-Hill.

and milling operations. In addition, the full application of CAM is more diverse than this relatively limited part production, encompassing a number of other manufacturing operations and controls, among which are NC verification, laser technology, automated nesting as shown in Figure 9.5, coordinate measuring machine for quality control, and the control of robots for loading and unloading, "pick and place" operations, welding, and spray painting, to name only a few applications of robot installations.

The flow of data from the CAD to the CAM operation is illustrated in simplified form in Figure 9.6, where the sequence of operations from conceptual design to manufacturing is shown primarily in terms of the development and transfer of design drawings and NC programs. The specifications are produced during the CAE/CAD operations, and result in sets of manufacturing drawings that constitute the controlling data for the fabrication activity, and the NC programming and other software required to run the machines and robots that create the final products.

The sequence of operations shown in Figure 9.6 demonstrates how the three operations are linked to each other, and the dependence of CAM on both CAD and CAE. It is important to recognize that the major differences between the CAD/CAM and non-CAD/CAM operations lie in the use of computers to generate the design and manufacturing data and not necessarily in changes in the operational procedures. Thus, there may still be significant human involvement and intervention, although primarily in the surveillance,

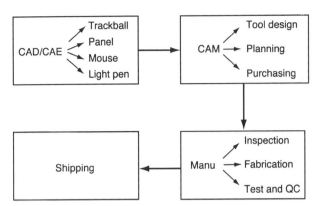

Figure 9.6   Diagram of data flow from CAD/CAE to CAM operations.

control, and test aspects of the CAM operation. The role of the electronic display system therefore remains in carrying out these activities, and the workstations used may be quite similar to those used for the CAD/CAE activities. However, it may be noted what when computer-integrated manufacturing (CIM) enters the picture, the role of humans is reduced to a minimal level, if not completely taken over by robots. One example is the virtual display shown in Figure 9.7. At this point, the display system function is reduced to a simple surveillance and may be further limited to some type of failure alarm, with terminals used only to determine the cause of the alarm.

As to the manufacturing workstations, they need not be as elaborate or advanced as those used for CAD/CAE, as may be seen from the parameter list shown later in Section 9.3. Indeed, the requirement for these workstations may be even less than for the minimal CAE workstation, depending on the extent to which the manufacturing operation is automated. In addition, it may be possible to operate with a minimum of personnel and associated workstations. In any event, it seems advisable, bearing in mind these considerations, that CIM is beyond the scope of this volume at present, and any consideration beyond these minimal comments must be deferred to some future edition or book. There are sufficient points of interest in the CAM application, particularly in its NC programming aspects and the concomitant electronic display requirements to warrant attention to CAM from the display point of view. To this end, it is of interest to list groups of specific manufactured parts and products that use CAM, listed in Table 9.2.

This is only a fairly limited list, and many other products and parts for assembly could be added, but the basic operations remain the same, and the need for terminal and display systems remains. The characteristics of the display system may range from the simplest terminal to the most complex workstation, depending on the complexity of the assembly and the amount of surveillance that is required to ensure that the product matches the

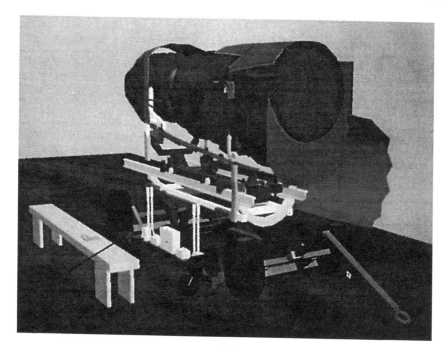

(*a*)

(*b*)

Figure 9.7 Example of a virtual image. Courtesy of McDonnell Douglas and Division Inc.

TABLE 9.2 CAM Outputs

Output	Process	Display
Wiring harness	EPATS[a]	X terminal
Quality control (QC)	Inspection and test	PC terminal
Plant	Construction	Workstation
Food and beverage	Control	Workstation
Medical implant	Manufacture	ASCII terminal
Automotive	Parts, body	Workstation
Aerospace	Parts, body	Workstation
Boards	X-ray inspection	X terminal
Fermentation	Respiration rate	PC terminal
Welding	Robotic	ASCII terminal

[a]Electrical planning and tooling system.

specification. Much of the display requirement is affected by the level of inspection, testing, and quality control involved, which in turn depends on the type of product. Thus, aerospace and automotive call for great care in these operations, and entertainment electronic products have somewhat less stringent demands. One example is the virtual display system used to model aircraft engine design for maintainability, as shown in Figure 9.7. Another is the display shown in Figure 9.8, which represents a food-and-beverage control system, and may be used for chemical, pharmaceutical, and other process industries. Each product type has its specific set of requirements that

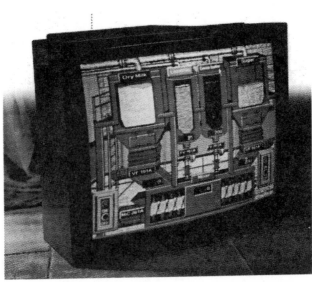

Figure 9.8 Example of drawing showing food and beverage control. Courtesy of AYDIN.

must be met before the product is released, and factors other than quality enter into the equation, in particular when the product is intended for the open consumer market, where competition is strong and quality may be subservient to cost. Of course, this is where the CAD/CAM operation adds significantly to the value of the product and has become the standard means for producing large numbers at reasonable cost.

9.3 SPECIFICATIONS

9.3.1 Introduction

As noted previously in Section 9.2.2, the specifications for the electronic display and hard-copy equipment and systems used for the CAD and CAE applications are similar to those shown in Table 6.6 for the VISIM aspect of these two applications. If CAM is added, these three applications may be grouped together because all of them are concerned with similar types of user applications and operating environments. Their actual requirements may differ to a considerable extent, ranging from the relative simplicity of CAM, at least insofar as the final outputs are concerned, to the maximum complexity that may be found in CAD. Thus, the types of systems that are best for each may differ as much as for the three categories of business applications or desktop publishing. Therefore, it is convenient to provide three sets of requirements, although the grouping consists of three separate applications rather then three aspects of single applications. This combination helps determine the choice of systems as well as the extent to which the same system may be used for more than one of this group of applications. This has the further advantage that users in the same company can readily move from one setup to another in an adjacent area. These combinations are shown in Table 9.3 for each of the three major applications. These results appear to be essentially identical to those for desktop publishing, which is true for the range of possible implementations. However, the specific performance capabilities required of the systems may differ even when the general system structure is the same. Therefore, it is necessary to examine the actual range of requirements for the operating application in order to arrive at a detailed specification. Thus, both CAD and the desktop publishing system that handles complex graphics are shown as needing high-end workstations,

TABLE 9.3 CAD/CAE/CAM

Application	System Requirements
CAD	High-end workstation
CAE	Server/X terminal, workstation
CAM	Server/ASCII terminal

but the actual applications involved may be satisfied by minimal versions of the general system category and cost savings achieved by going into more detail. However, this amount of detail is beyond the intentions of this text, and is left to the user, following the outlined procedure and avoiding the need to examine all possibilities. The same may be said of most of the applications covered here.

9.3.2   CAD/CAE

9.3.2.1   *Display Specifications*

As noted, the dynamic display specifications for these two applications are quite similar to the ones shown in Table 6.6a. The CAD/CAE versions of this table, given in Table 9.4, may be compared with those given in a previous table to note the differences. It is not done here in the interest of conciseness, and because the differences are relatively small.

The CAD application does call for a fairly high level of dynamic electronic display performance because of the potential complexity of the images needed to carry out many of the design tasks. This range of complexity is indicated in Figures 9.9 and 9.10, and even more complexity is possible with the new high-resolution monitors available in both CRT and LC FPD technologies. The types of applications encountered with CAD/CAE range from designing and generating the simplest circuit schematic of the type shown in Figure 9.11, or the more complex piping schematic shown in Figure 9.12, to a highly complex mechanical drawing of a group of castings. More recently, virtual reality has come into the picture, in conjunction with 3D to make it possible to design and generate models that simulate the actual product to be manufactured, thus making the transition to the final fabrication simpler and much less prone to errors. The advantages of using CAD/CAE, with their concomitant CAM, have become quite apparent, as is becoming the case for CIM as well, although the last has much lesser significance for electronic display applications.

TABLE 9.4   Processor and Dynamic Display Specification for CAD/CAE Applications

Parameter	Value	
	CAD	CAE
Luminance (cd/m^2)	50–300	50–150
Contrast ratio (number)	10–40	10–30
Resolution (TV lines)	800×600–1600×1200	800×600–1280×1024
Colors (no.)	256–16 million	256–16 million
Viewing area (diagonal in.)	17–32	15–27
Speed (MHz)	15–100	100–540
Memory (Mbytes)	66–100	120–540

Figure 9.9 Example of complex imagery. Permission to reproduce granted by Hewlett-Packard Co. Copyright Hewlett-Packard Co.

The hard-copy display specification for the CAD/CAE application is similar to Table 6.6b, and is given in Table 9.5 so that the relatively small differences can be indicated. The reason for the differences is that the full CAD/CAE application is considered rather than the VISIM aspects alone. This is a distinction without much differences as seen from the values given in Table 9.5.

The drawings produced that meet this specification are similar to the images shown in the previous three figures, and the requirements for the hard-copy specification are clearly quite similar to those given for the dynamic display, and usually somewhat more severe. It should also be noted, as is indicated by the display examples presented previously, that a wide range of visual requirements are possible when all of the many different design and engineering activities that fall within the general CAD/CAE applications are considered. This can and has led to a variety of programs and operational procedures, as may be determined by examining the host of different programs offered and the complexity of the resulting designs and

Figure 9.10 Example of complex imagery. Permission to reproduce granted by Hewlett-Packard Co. Copyright Hewlett-Packard Co.

analyses that may result. In any event, it is clear that CAD/CAE is an important application for electronic displays and has significant impact on the quality and performance capabilities demanded of the display equipment and systems.

9.3.3   CAM

Next in the sequence of computer-aided operations from initial concept to manufactured product comes the CAM or manufacturing part of the total operation. The results of this activity must justify the effort and cost of both the two previously examined steps, as well as this final step. It might be of some interest to note that it has taken over 40 years for CAD/CAE and CAM/CIM to reach their present state of capability and acceptance, and that there were many problems to be solved along the way from the early efforts at MIT, IBM, Convair, and GM to the present highly sophisticated systems and designs resulting from the presently available equipment, systems, and programs designed for use in computer-aided operations. The

Figure 9.11   Example of a schematic diagram. Courtesy of CalComp.

computer-aided sequence, now termed CAD/CAE/CAM/CIM, has taken over to a considerable extent almost all of the design and manufacturing operations concerned with the manufacture of a wide range of articles and products enjoyed by the consuming public at reasonable cost and good quality. The role of electronic displays in the CAM operation is not quite as demanding as for the CAD/CAE activities, but it is still quite significant, and calls for a high level of performance in some of the applications, such as those listed in Table 9.2. This level of performance is represented by the parameter values shown in Table 9.6. These are relatively undemanding specifications, but should be adequate for the majority of the CAD application groups. However, as the cost of electronic display equipment and systems comes down, it may be economically feasible to obtain better performance for lesser expenditures, so that the performance demanded by the CAD/CAE applications groups may also be obtained for the CAM operations. At the other extreme, as CIM becomes more prevalent, the display requirements may be considerably reduced, or totally eliminated.

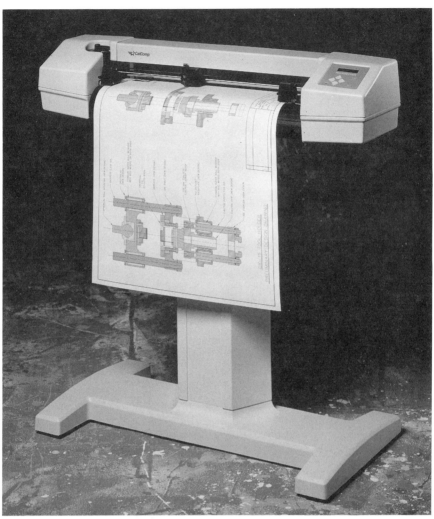

Figure 9.12 Example of a complex mechanical drawing. Photo courtesy of CalComp.

TABLE 9.5 Hard-Copy Display Specification for CAD/CAE Applications

Parameter	Value	
	CAD	CAE
Resolution (dpi)	200–1000	200–500
Speed (ppm)	1–3	1–2
Colors (no.)	4000–16 million	256–4000
Size (in.)	8.5 × 11–21 × 40	8.5 × 11–18 × 27

TABLE 9.6   Processor and Dynamic Display Specification for CAM Application

Parameter	Value
Luminance (cd/m²)	25–50
Contrast ratio (no.)	5–15
Resolution (TV lines)	640 × 480–1024 × 800
Colors (no.)	16–256
Viewing area (diagonal in.)	13–25
Matrix ($W \times H$)	600 × 400–1000 × 800
Speed (MHz)	8–16
Memory (Mbytes)	40–120

However, this situation is still rather far in the future, so electronic displays for CAM should remain of significance for some time to come, at least until the next edition of this text is published. Meanwhile, for now and into the foreseeable future at present, the requirements for CAM should fall within the limits listed in Table 9.5.

9.4   BLOCK DIAGRAMS

9.4.1   Introduction

The block diagrams presented next are for the specific system configurations that may be used to produce the images required for the proper operation and use of the electronic display systems associated with the CAD/CAE and CAM applications discussed in the previous sections of this chapter. How-

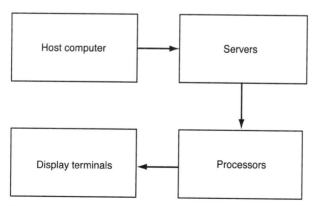

Figure 9.13   Generalized CAD/CAE/CAM block diagram.

ever, prior to presenting the individual block diagrams for the systems used for each application, it is useful to show the common elements among them by means of the diagram shown in Figure 9.13. This simplifies the discussion of diagrams for these applications by demonstrating the similarities, as well as providing an introduction to the differences among the block diagrams found in the following section devoted specifically to the individual application groups.

Figure 9.13 begins with the *host computer*, which refers to the computer or computers that perform the basic operations required to carry out the tasks determined by the applications involved, and may be somewhat similar for all three, although the NC machines are most common for CAM and most of the graphics workstations contain their own computers. These differences are shown in later sections. Next, following and controlled by the host computer system, are the servers devoted specifically to the application under consideration, again with the proviso that the workstation may include or supersede the server. Following the individual server may be separate graphics processors that are devoted primarily to producing the digital data required to generate and control the imagery shown in the terminals. Then, finally there are the different types of display terminals, ranging from the simplest ASCII units to the highly sophisticated graphics workstations. These terminals are quite varied in type and complexity, and are probably the most important part of the total system insofar as the electronic display portions are concerned. All are considered separately in the following sections, and expanded as necessary to demonstrate the differences among the requirements for the three basic applications covered in this chapter.

9.4.2   CAD

As the first among the applications covered in this chapter, the block diagram for the CAD application contains all the basic elements to be found in the

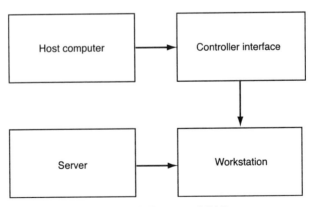

Figure 9.14   Block diagram of CAD system.

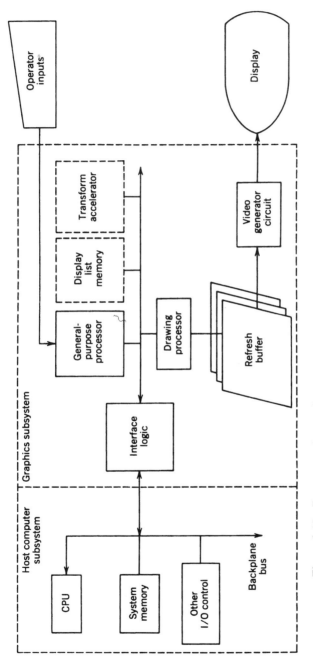

Figure 9.15 Representative workstation block diagram. After Breen [3], by permission of SID.

other two block diagrams that apply to CAE and CAD applications. As such, it is necessary to introduce some of the basic aspects of the choice of the elements shown in Figure 9.14 in order to clarify the reasons for the choices.

The block diagram in Figure 9.15 differs from that shown in Figure 9.13 only in that it can be operated with only a server in addition to the workstation because the workstation contains sufficient data processing and memory capabilities to render the host mainframe computer unnecessary. This capability is shown in the workstation expanded block diagram presented in Figure 9.15, which contains all the elements that are needed to constitute a fully instrumented graphics workstation. It is the one shown in Figure 4.3, related here to the CAD requirements. It should also be noted that in some cases the workstation alone may be sufficient to meet the specific application requirements. In addition, the CAD application usually requires some type of hard-copy output device as part of the display block, so that all the data output can be made available in hard-copy form.

9.4.3 CAE

As noted previously, the CAE application usually has somewhat lesser requirements in regard to the electronic display portion, although the data processing requirements may be at least as severe as, and in some cases more demanding than, those for the CAD application. This leads to two different configurations, one of which may be identical to the server–workstation part of the CAD block diagram, whereas the other may differ from that diagram in that more of the data processing capability is contained in a host computer, and the display may be contained in a simpler terminal such as the X terminal shown in Figure 4.4. In either case, the electronic display may be simpler than that required for the CAD application. In any event, it is useful to show another block diagram that could fit the lesser requirements from the display point of view of the CAE application, and this is shown in Figure 9.16. In this diagram, somewhat more attention is paid to the processing and memory portions of the unit, whether it is a workstation or X terminal, and the monitor may be of either the CRT or FPD variety, with the appropriate operating characteristics as listed in Table 9.4.

9.4.4 CAM

The CAM electronic display requirement is in many ways the least demanding of the CAD/CAD/CAM application groups, although in many respects it and the associated system, terminal, or workstation equipment have much in common with those that meet the operational requirements for the other two applications. Thus, any block diagram for the requisite system will be quite similar to the ones shown in Figures 9.15 and 9.16, and it is not necessary to repeat them at this point. However, it is appropriate to emphasize that the differences in the electronic display requirements from those of

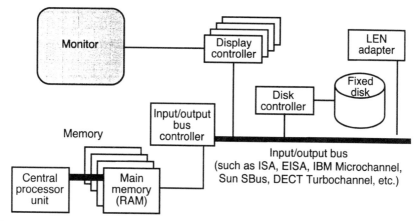

Figure 9.16   Block diagram of a CAE workstation. After Machover [4], by permission of McGraw-Hill.

the other two, as shown in Table 9.5, are not trivial, and may result in significant cost reductions that could warrant issuing specifications for the CAM applications separate from those issued for the CAD and/or the CAE applications. This choice must depend on the specific type of manufacturing activity that is involved, and it is not possible to make recommendation at this point beyond those included in the tables in this chapter that relate to the CAM activity in general. Again, it should be noted that as the costs of the different configurations and display equipments come closer together, it may be appropriate to treat all these requirements for the three applications as essentially synonymous, but this condition, desirable as it may be, remains in the as yet unpredictable future.

9.5   CONCLUSION

The CAD/CAE/CAM group of electronic display applications constitute a highly significant part of the total display application area, and contribute importantly to the total market for computer-aided operations. Therefore, they must be carefully considered and their requirements fully evaluated when vendors of this type of equipment try to determine what the performance characteristics of the products to be offered should be. High on the list are the graphics workstations that have become the workhorse of computer-aided applications, with the X terminals and the simpler ASCII terminals as lesser but still important components of total operating systems. Therefore, the information contained in this chapter should be of use to both end users and product venders concerned with the application areas covered in this chapter. To this end, close attention to the information contained in the discussions, block diagrams, and parameter table should contribute useful data and procedures that may be used to facilitate and simplify the task of arriving at the optimum combination of design, analysis, and manufacture required to achieve the most acceptable results in terms of product cost, reliability, and operational adequacy.

REFERENCES

1. Machover, C., Ed., *CAD/CAM Handbook*, McGraw-Hill, New York, 1996, Section 5.2, 33.
2. Machover, C., Ed., *The CH Handbook*, McGraw-Hill, New York, 1989, Section 5, 220.
3. Breen, P. T., "Review of Workstation Technology," *SID Sem. Lect. Notes* 1991, **M9**/11.
4. Machover, C., Ed., *CAD/CAM Handbook*, McGraw-Hill, New York, 1996, Section 36.1, 486.

10

VIRTUAL REALITY

10.1 INTRODUCTION

Virtual reality is a group of applications that have received much attention recently, with specific examples ranging from medical applications to entertainment. Before defining the term *virtual reality* (VR), it is of interest to provide a list of the most important examples of applications in this group, as a preliminary to the more detailed explanation to follow. This examination of the specific applications that use some form of VR implementation establishes a base for the discussions that follow. This list is given in Table 10.1. This is only a representative list, and many other applications are either using or exploring the use of VR for various purposes. In addition, a large number of manufacturers, especially in Japan, are developing and marketing VR devices and systems. These are primarily a variety of head- or helmet-mounted displays for VR that go by the names listed in Table 10.1 in the "Displays" column, to which are added a number of system components to make a full stereo VR system. These latter elements are also shown in Table 10.1 in the final column, some of which have commonly accepted names, and others have names given the specialized functions required by the vendor. The meaning of the designations and the technologies involved are covered in some detail later, and it suffices to state at this point that the basic display unit consists of some type of stereo viewing glasses, connected to a driver engine that supplies the signals required to operate the viewing element in synchrony with what the viewer expects to see. System descriptions are also included in the technical discussions found later in order to more fully understand the total requirement for a working VR system, although the main interest here is in the display.

TABLE 10.1   VR Applications

Application	Displays	Other VR Units
Advertising	Head-mounted stereo (HMS)	Desktop computer
Architecture	Head-mounted display (HMD)	Desktop computer
CAD/CAM	HMD	Workstation
Education	3D glasses	Desktop computer
Entertainment	3D glasses	VR entertainment system
Games	3D glasses	Desktop computer
Manufacturing	HUDSET	3D tracker transmitter
Medical imaging	HMS	DataGlove, viewing wand
Simulation	SIM EYE	Reality engine
Television	Stereo visor	Stereo driver

10.2   APPLICATION DESCRIPTIONS

10.2.1   Introduction

Virtual reality (VR) is the term in use for a group of applications that use the visual characteristics that create the general impression for the viewer of experiencing a real-life activity. However, the actual applications that employ these visual and psychological effects can vary by wide margins from each other, as is evidenced by the list given in Table 10.1. It is the purpose of the descriptions given in this section to clarify the manner in which the displays are used to create the visual illusions, and how the application of VR to the specific operation assists in the performance of the tasks involved. The discussions that follow are intended to achieve this result on an application-by-application basis, in alphabetical order. This requires more detail than if only VR as a unique application in itself were covered, but does allow the special attributes of VR displays to be clarified. However, prior to embarking on these descriptions, it is advisable to describe how VR operates in a generalized way so that its use in these specific applications can be better understood.

Virtual reality displays are essentially extensions of the three-dimensional displays that have allowed stereo vision to be applied to the viewing of scenes and objects in a way that simulates the real world. The various technologies that have been developed to allow such stereo images to be generated and presented on some type of screen are reviewed later in this chapter, but at this point it is sufficient to note that all these technologies use some means for providing spatially separate views of a scene to each eye separately. The problem is to create these two images in the proper relationship, and then present them in a manner that allows one eye to view only its own image in conjunction with the other eye so that the stereo imaging replicates what the eyes normally see in real life. The optical techniques go back to the nineteenth century, when viewing static stereo images by means of some device such as the one termed the *Stereoopticon* was a popular means of diversion and entertainment at home prior to the advent of television. This entertainment application was quite successful in its time, and various techniques for viewing cinematic renderings of mobile images have been tried out, in theaters and in the home, usually with minimal success and very limited acceptance. The difficulty has usually been the need for special viewing implements, such as colored spectacles in which each eye has its own lens color and the two stereo images are presented in the color corresponding to the eye lens color. This approach to dynamic stereo displays can create fairly acceptable stereo scenes, but requires rather restrictive physical limitation to be convincing. For example, the image must be viewed head-on for best results. This limitation has led to rather unsuccessful results, but modern systems are overcoming these limitations, in particular the use of special vision restricting devices such as head-mounted displays (HMDs) and various versions of 3D stereo. The technical details of these units are discussed later, and at this point it is sufficient to state that the displays, in conjunction with eye and head tracking systems, have resulted in highly effective VR systems and led to effective applications of the technique as described in the following examples.

10.2.2 Advertising

Advertising that employs dynamic displays to convey a message and/or influence the viewer to follow a particular pattern of activity is an ideal example of how VR displays can be used to achieve a stated goal. The need for the special viewing devices has been something of a limitation, but this has been overcome to some extent by setting up a group of viewing stations coupled with the appropriate viewing devices. One example of this arrangement is the "virtual brewery," set up at Sapporo headquarters in Japan. The exhibit uses what is termed a *fixed stereoscopic viewer* (FSV), which consists of a dual-view display, and the need for a fixed view is overcome by having a separate navigator unit that can be used in conjunction with a number of FSVs. The result is an effective trip through the brewery for a number of individuals at the same time. Thus, the installation becomes an effective advertising tool for the brewery. Another advertising-related use is for automobiles where a customer could essentially operate a vehicle while sitting in an instrumented automobile body, instrumented for VR, and wearing an HMD. Other similar uses are possible for homes and buildings. An even more striking example of the use of VR for advertising purposes is one that Exxon has used in an advertising campaign where a car is converted (morphed) into a tiger.

10.2.3   Architecture

Architecture is an application that can benefit immensely from VR, in a manner somewhat similar to its use in advertising noted previously. Essentially, the procedure is to use one of the special stereo viewers or an HMD setup for the viewer to see the architectural object in a VR mode. The combination of a head tracking unit, a computer with an appropriate VR program, and the HMD for viewing the resultant displays can create a very realistic portrayal of the architectural object, changing as the viewer moves through or around it. This capability is of use to the architect in determining without physical models what the appearance of the designed object will be, and assessing the effect of changes on the final result. In addition, a prospective customer can see a simulacrum of the final result of the design, and judge whether it is satisfactory. Indeed, the viewer can walk through every room of a house, trying out different colors and other features, before deciding whether the potential result is likely to be satisfactory. The advantages of this approach to both the architect and the customer are incalculable, and this application may become one of the most important.

10.2.4   CAD / CAM

This is a somewhat all-encompassing application in that it includes both the initial design aspects and final manufacturing process involved in producing a final product. The question that arises when VR is studied as a possible capability for use on these types of activities is whether the VR aspects lend anything significant to the total process aside from the novelty of depicting what the results will be before anything is built. The possibilities appear most attractive in the design and manufacture of large objects such as airframes and automobile bodies where the need for full-size mockups may be eliminated with a significant cost saving and improved designs. In addition, the assembly of a large number of parts may be viewed in 3D to determine whether there are any space conflicts. Finally, the actual manufacturing process may be improved by accurately locating in 3D where certain operations such as drilling should be performed. The question always remains whether the cost of the equipment is greater than the savings due to eliminating certain steps. Of course, the same arguments were presented when CAD was first proposed, and it took many years before the accountants were convinced. However, this background of experience with new approaches may have made the acceptance of such radical changes in the design process more likely. It might be noted that a somewhat simpler approach than true VR is being investigated by a group at Boeing [1] and may lead to greater acceptance of VR-type of approaches. In any event, the possibilities are manifold, and much may be expected of the performance of VR systems, especially if the displays improve and costs are reduced. The performance aspects are discussed later, but it might be noted here that the inadequacies consist largely of low resolution and luminance, and large heavy headsets. However, progress in overcoming these limitations should continue, and the use of VR in CAD/CAM increase rapidly in the not-too-distant future.

10.2.5 Education

VR techniques can benefit students at any level, from kindergarten to graduate school. It exposes young children to experiences that would be impossible in the classroom such as visits to museums, zoos, and aquariums and other beneficial activities. Of course, this artificial world must be supported by experience with the real world, but the educational opportunities are greatly enhanced by prior and supplemental exposure to live examples. Moving further along the educational chain, science teaching can be much improved by the use of VR experiences that show the scientific images is their full 3D form. This approach can be particularly revealing in such sciences as biology, chemistry, and physics, where the ability to view DNA, molecules, and atoms can bring apparent reality to what have been represented by models that the student must take on faith. Many other disciplines such as astronomy, mathematics, and even music and visual arts can be made vivid by the use of VR. The equipment is available to at least begin this use of VR, and the improvements that will surely come about should make VR an indispensable tool for educational applications.

10.2.6 Entertainment

Entertainment is a widely varied application, ranging from the simplest handheld game to complex 3D presentation of involved cartoons and expeditions into historic sites as well as 3D live television. Games and TV are sufficiently important to warrant separate treatment, and they are covered later in this section. However, entertainment as a general topic has enough potential to be afforded its own section. Although the examples of this application are still few and new ones are still in the infancy stage, it is possible to draw some conclusions as to its probable future. One interesting prospect is the expansion of CD-ROM presentations of museum artwork on the computer screen into full 3D images that can be viewed directly on the screen by means of 3D stereo or HMDs, with the further option of actually walking around a gallery and seeing all the art objects in their proper setting. Another possibility that is being explored is the development of a virtual museum to supplement and expand the CD-ROM versions, where the artwork and the museum are available through modem connections as in on-line services. This type of application also has educational implications, but in this discussion it is considered as a source of pleasurable experience also sometimes known as "entertainment." Carnegie Mellon University is one location where this type of VR development is occurring. However, one important desideratum is that the VR experience must offer the display quality that enhances the experience, and this requires that there be considerable improvement in the performance capabilities of the display. This means higher resolution, image size, and color quality, all of which are certainly on the way, if not actually in production at present. This is indicated by some of the data on equipment performance provided later. At this point it suffices to say that entertainment is an important application for VR in both the near and distant future.

10.2.7   Games

This application might be thought of as a subgroup of the entertainment application, and included in that discussion. However, it is so pervasive and important that it warrants treatment as a separate item, with the overlap between it and entertainment understood but not belabored at this point. The advantage of this approach is that it allows the unique aspects and features of the games application to be given separate and more complete attention. First, as a general definition, games are considered here as any activity whose main goal is the involvement of one or more individuals in some defined series of steps whose outcome is essentially self-contained within the game. Of course, with the intense involvement of many individuals as fans in the outcome of many games, it might be said that some games are more important to some individuals than are their more social day-to-day activities, but this need not alter the general definition. Given this broad definition, the range is so varied that simple games such as solitaire, uncommon ones such as tiddlywinks, and highly complex ones such as "go" or 3D chess, all fall under this rubric. However, the main attention is given to electronic TV and handheld games, where the quality of the visual display can play an important role, and the VR aspect may add significant interest to the playing experience. This is particularly effective in interactive group games where the individual experience can be significantly heightened by the VR illusion. The most prominent examples of VR games are those located at arcades such as Virtuality and Mirage, to name only a few. In the first the player flies through simulated battle scenes and engages with fantasized enemies, responding to attacks and other events. The latter is based on a training technology, where the scenario has "Star Wars" experiences for pairs of players. These and similar games are becoming available for home use, and other installations may be found in museums and schools in the future. The possibilities are enormous and are limited primarily by the ability of the vendors to design new scenarios. Also, some permanent installations of fully equipped chairs capable of playing many different types of games should be appearing at arcades and malls, although the installation costs may be quite high. In any event, VR has important contributions to make to games experiences and probably will be a driving force as to the direction development takes, replacing military applications in this role to a considerable extent.

10.2.8   Manufacturing

A number of manufacturing operations may benefit from the use of VR beyond those involved in CAM. Some examples are the ability to see a location in 3D, using VR techniques, so that a drill position can be accurately determined, and the use of a "virtual lathe," where the object to be machined appears as a 3D VR image in space, so that the remote operations using a mouse or data panel to control the lathe can be combined with a realistic view of the object being machined. High-resolution systems and head motion sensors with six degrees of freedom make this a powerful tool. When such capabilities in manufacturing are coupled with the design and modeling abilities made available by the use of VR for CAD/CAM, the combination provides important improvements in manufacturing cost, design time, and final product quality.

10.2.9   Medical Imaging

There are a surprisingly large number of medical applications for VR, one of the most important of which is imaging. This capability is of value not only in medical education and training but also as an operational tool in radiation applications. It enables the physician to see more precisely what is happening than would be possible with a standard presentation. The patient's body is created, in 3D, by a CAT scan or some other means of scanning the body, such as MRI. Then by wearing an HMD the physician can see a VR image of the tumor site, and direct a beam of radiation directly to the tumor. Surgical procedures may also be expedited by similar VR approaches. Of course, the quality of the display is of major importance in determining how well the tumor can be localized and the radiation beam directed. High resolution is of major importance, as is color fidelity. Another important medical application of VR is in education and training of both physicians and lay students. Several software programs are available that make it feasible to see the human body in all its natural complexity and delve into it organ by organ. In addition, it is possible for the medical student to learn how to perform operations by following the surgical procedures on the VR model. Other uses for VR in medical applications are rapidly developing, and their success and use will depend on easde of use, and how realistic the VR model is.

10.2.10   Simulation

Simulation is an application that is most suited for VR, with its need to approach as close to reality as possible. Simulation systems have been in use for many years as a means to provide the visual and operational environment for training in the use of complex control systems, especially in aircraft and naval vessels. The systems presently in use are large and complex, and the VR approach makes it possible to considerably simplify the training system and achieve better results in the training simulation. Of course, the same considerations also hold for games where battles are simulated and the player responds to the simulated situation. However, the discussion here concentrates on the more practical aspects of simulation as an application for VR.

One important example of the use of simulation is in pilot training, where the capability of subjecting the present or future pilot to the problems that will arise when the aircraft is under the individual's control is of prime importance in training and education. This is particularly evident when a new aircraft is unveiled, and even seasoned pilots need to undergo a training course on the idiosyncrasies of the aircraft. This has been done by means of complex simulators where the entire inside of the cockpit is built and control mechanisms installed that cause the visual image to follow what the pilot should experience. These simulators are quite expensive but have proved to be well worth the expense. Now this cost can be significantly diminished and the training experience made more realistic by means of VR. By using an HMD in conjunction with computers and software that simulate the aircraft control experience, the visual image can be made to approximate what the pilot would see under actual flight conditions in the real world. Similar techniques are used in aircraft games that simulate battles with considerable success, and the use of VR simulators will surely repeat this result. The same considerations hold for other navigation and training situations, where large installation now in use can be replaced by reasonable-sized VR units.

These applications impose rather severe requirements on the visual display equipment, particularly when the results must be compared with the highly successful simulators that are presently in use. Anyone who has experienced the high level of realism achieved in some naval simulators must realize that the VR display meets a very high standard of performance. The maximum in performance must be achieved in resolution, color accuracy, speed of response, and general realism to be acceptable. This type of performance appears to be within reach in the foreseeable future.

10.2.11 Television

Television is perhaps the most promising application for VR of all listed in Table 10.1, at least as far as its possible commercial success is concerned. This is particularly true because of the imminent advent of high-definition TV (HDTV) in the near future. Of course, TV could be included under the general rubric of entertainment, but it is so large and important on its own that treating it separately is warranted. There have been some experimental attempts to achieve 3D TV with only limited success, and the same might be said of cinematic efforts, but new achievements in the latter have stimulated interest in the former even more than the still limited general public demand might warrant. However, it has been established by many experiences in innovative approaches to these types of capabilities that the demand will grow rapidly if the material offers interesting experiences. This certainly may be said of the VR results that have been achieved in a number of the other application areas covered previously, and it may be anticipated that the same results will be achieved once the techniques and equipment have been brought to the stage of relatively low-cost and high-quality results. Although the source material would have to be developed by the broadcaster or special program developers, this would be no different from what is presently the standard procedure. Alternatively, videos for the various types of players available for showing videos on TV screens could be produced and played on standard sets, in association with some form of HMD or other 3D viewer. Thus, TV and its associated VCRs, DVDs, and laser players can extend the use of VR to this general entertainment category.

This completes the short reviews of a small group of applications for which VR appears to be well suited. There are many others not mentioned here, but the possibilities appear to be limitless and the future is not predictable if the technical performance requirements can be met. This question is covered in the next section.

10.3 TECHNOLOGIES

10.3.1 Systems and Software

A group of possible elements that might make up a VR system are shown in pseudopictorial form in Figure 10.1a. This illustration includes a number of elements in addition to those concerned specifically with the display, such as the instrumented gloves, speech, stereo sound, and various tools that might be used in such a system. The result is a highly complex configuration, of which the display and its associated elements are only a part. A basic system block diagram in its simplest form for one type of VR configuration that highlights the display portion is shown in Figure 10.1b.

The VR configuration shown in Figure 10.1b consists of a computer that can be programmed to generate the proper right and left images of whatever the VR scene might be, some type of HMD or other stereo viewer, and the tracker system that measures eye movement and transmits the resultant error signal to the computer so that the proper corrections may be made to the images to generate the appropriate changes in the visual scene. Of these items of equipment, the ones of prime importance for display applications are the viewers that enable the user to see the VR imagery in proper form and the tracker that develops the signals that define the viewing direction. These signals are sent to the computer, where the program generates the new signals for the display to provide the proper new images. The result can be a highly realistic set of active images that can convince the viewer of the reality of the scene. Although the display is shown as contained in a helmet, other forms of 3D displays may also be used if there is some means included for tracking either eye or head movements. An example of a VR sequence that

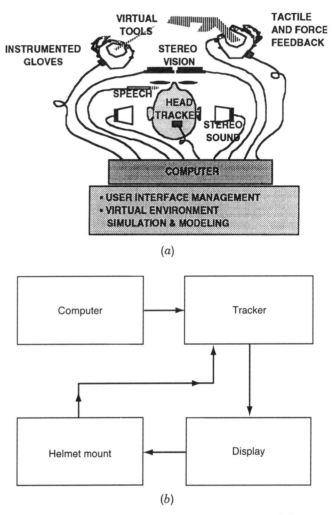

Figure 10.1 (a) Elements of a virtual reality system. After Baum [1], by permission of SID. (b) Block diagram of a virtual reality system.

might be generated by this system in association with appropriate software is shown in Figure 10.2 and represents one sequence of images that might be created. It was created by Adams Consulting (Chicago, IL), using Superscape software (Santa Clara, CA).

Another system, built around an HMD and intended to integrate the HMD into a tactical aircraft, is shown in Figure 10.3. The basic elements shown in Figure 10.1a are expanded to show the specific tracker sources and processor, as well as the display processor, CRT drive, and pilot controls that affect the display image.

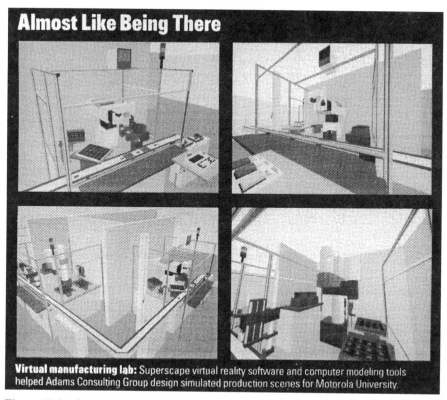

Almost Like Being There

Virtual manufacturing lab: Superscape virtual reality software and computer modeling tools helped Adams Consulting Group design simulated production scenes for Motorola University.

Figure 10.2   Sequence of virtual reality operations. Image shown was created using Superscape software (Santa Clara, CA) by Adams Consulting (Chicago, IL).

A somewhat more complex VR system is pictured in Figure 10.4. This is an advanced experimental system that operates as a virtual HUD, with the capability of writing imagery directly onto the retina from the display surface that can appear to be from 9 in. to 20 ft. in front of the viewer, and is superimposed on the real world. This is a highly imaginative approach to a VR type of display system, and has numerous applications besides the battlefield situation for which it was originally intended. It may be the interactive VR system of the future where sensor and user inputs are combined, and a holographic image is produced that appears to the user to be at a range of positions in space.

The procedure involved in producing the VR world imagery from a software point of view might be represented by the block diagram shown in Figure 10.5, where the 3D objects required for the VR scenes are created and stored, along with other images as well as real-world inputs. The objects and images stored in the files may have a good deal of complexity, and the ability of the display to present them effectively depends on the performance

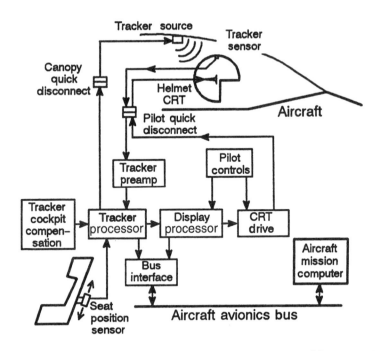

Figure 10.3 Typical helmet-mounted display system. After King [2], by permission of SID.

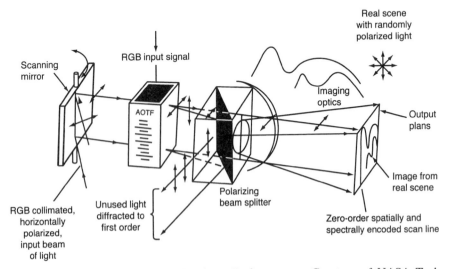

Figure 10.4 Diagram of virtual head-up display system. Courtesy of NASA Tech Briefs.

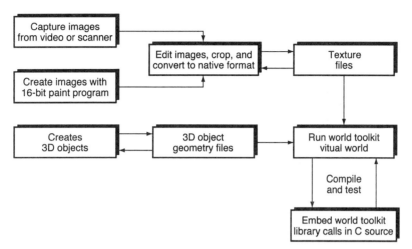

Figure 10.5   Block diagram of virtual reality world imagery. Courtesy of *Computer Design* (Nov. 1994, p. 56).

capability of the display system, whether it is an HMD or other type of 3D viewer. The capability requirement is covered in detail in the next section. In any event, this type of software enables the developer to create a variety of virtual worlds that can be used for a wide range of specific applications. Numerous companies specialize in producing software as well as complete scenarios for VR presentations, many of whom have home pages on the World Wide Web, so there is no difficulty in accessing these sources. Therefore, it is possible to obtain the necessary software, as well as hardware material for many of the possible VR applications listed in Table 10.1. This includes software for the trackers for head and/or eye motion, an instrumented glove, other hand-operated input devices, sound, and motion. These are the problems that the VR developer must face in preparing an operational system that can drive the various elements of the VR hardware used in the application. This hardware must include a general-purpose hardware platform of the Pentium or high-speed Unix variety, and the software, in turn, must be designed to provide the necessary tools to expedite the operation of such systems. This results in as many as five different architectures that may be used to implement the system for a VR application, with the direction of information flow shown in Figure 10.6; briefly, this information flow is as follows:

View information scene database → objects → screen primitives → image

This table shows the five different approaches that may be used as processes for VR applications. Clearly, type 1 is the simplest and least expensive, whereas type 5 is at the other extreme. The host processor may be of the

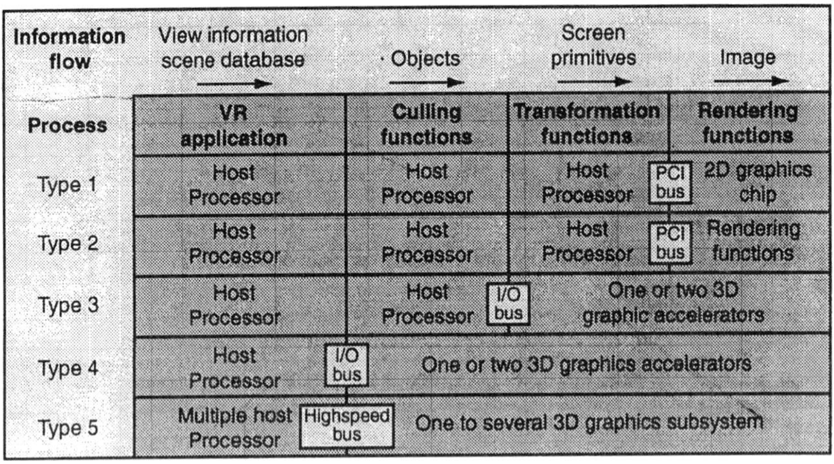

Information flow	View information scene database →	· Objects →	Screen primitives →	Image →
Process	**VR application**	**Culling functions**	**Transformation functions**	**Rendering functions**
Type 1	Host Processor	Host Processor	Host Processor [PCI bus]	2D graphics chip
Type 2	Host Processor	Host Processor	Host Processor [PCI bus]	Rendering functions
Type 3	Host Processor	Host Processor [I/O bus]	One or two 3D graphic accelerators	
Type 4	Host Processor [I/O bus]	One or two 3D graphics accelerators		
Type 5	Multiple host Processor [Highspeed bus]	One to several 3D graphics subsystem		

Figure 10.6 Diagram of virtual reality information flow. Courtesy of *Computer Design* (Nov. 1994, p. 66).

Pentium or Power PC types, and perform most of the functions as is shown in Figure 10.6 for the types 1 and 2 systems. In the case of the types 3 and 4, the host processor is supplemented by 3D graphics accelerators to perform all the functions, whereas in type 5 at least one full-graphics subsystem with multiple host processors is required. The range of system possibilities for the various different VR configurations is quite large, and the resultant capabilities are similarly varied. The same considerations apply to the actual equipment that is used to make up a VR system, and these are covered in the next section.

10.3.2 VR Equipment

10.3.2.1 Host Processor

As noted above, the host processor may be a standard, high-performance PC when only relatively limited processing capability is required, as is the case for some of the simpler advertising and educational applications. However, it may also be necessary to expand this capability to meet the more elaborate requirements of the CAD/CAM, medical imaging, and simulation applications by using more elaborate units such as minis and servers, or multiple combinations of PCs. Indeed, with the apparent resurgence of interest in using mainframes, there may be a place for these types of computers as well. However, at this point, the discussion is limited to the first three possibilities, and the performance capabilities of each type, as related to that of the total VR system, are listed in Table 10.2. The parameters of prime interest are processing speed, database memory size, bus type, and I/O capabilities. A number of units offered by various vendors meet the needs of a variety of VR

TABLE 10.2   Processor Parameters

Unit	Type	Speed	Number	Size	Bus
Processor	Pentium	90–120 MHz	—	—	—
Multiprocessing	—	—	2–4 units	—	—
Standard memory	—	—	—	16–201 (Mbytes)	—
Hard disk	—	—	—	420 Mbytes–2.1 Gbytes	—
Video	—	—	—	—	PCI

Source: Fernie [4].

systems and applications, but the parameter values shown in Table 10.2 are only representative and not tied to any one of the available systems. These are only representative values from one source, and wider ranges may be available if required. In addition, lower-capability desktop units may also be adequate for some VR applications. However, it should be recognized that with the rapid growth of applications, it is most likely that the processors with maximum capabilities will be preferred.

10.3.2.2   *Helmet-Mounted Display (HMD)*

HMDs are probably the most effective and popular form of display in use for VR applications. They are an outgrowth of the head-up display (HUD) in common use in aircraft as a navigation aid. This display overlays part of the instrument panel on the pilot's visual field or provides a simulated landing image of the type shown in Figure 10.7 that repeats Figure 4.15. The advantage of the HMD for this type of application is that the image on the screen or windshield can follow the motion of the user's head so that the generated image is put in a proper relationship with the real world that can be seen through the windshield. The same technique has been tried out in automobiles with some success, but the HMD has shown particular capabilities for VR applications and is used in a number of simulations and other real-world situations.

Several technologies have been used for HMDs, and one that has been quite successful used two liquid crystal light valves (LCLVs), one for each eye. The stereo imagery is generated by the computer system, displayed on the two LCLVs in appropriate synchrony and spatial relationship, and transferred to the user by means of helmet mounted optics. The imagery is controlled by the head movement through the tracker, and therefore follows or simulates the actual changes in the viewed image. One example of a possible diagram for a virtual HUD system is illustrated in Figure 10.8, where a color capability is included by means of multiplexing three color signals into an acoustooptic tunable filter (AOTF) in order to form spectrally and spatially encoded scan lines. A pulse source illuminates the filter periodically, and each scan line enters the cell aperture, and an oscillating scanning mirror maps each scan line into its correct position on the output plane. A real-world

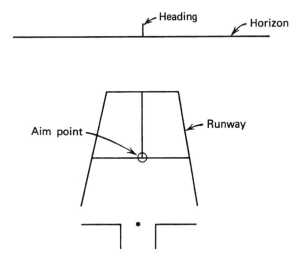

Figure 10.7   Example of simulated aircraft landing image.

scene may also be imaged on the output plane and be combined with the VR-generated image. Although this design represents only one approach and is relatively complex, it does show what may be done by using an HMD as the display means for a VR system. It or other similar systems may be extended further into a retinal display by means of the optics shown in Figure 10.9.

It is also of interest to list the display parameters and their values as related to the HMD. These are shown in Table 10.3 for two military units.

The HMDs have an image collimation system in front of the helmet, where the 40° unit uses a catadioptric collimator, and the 60° units have a

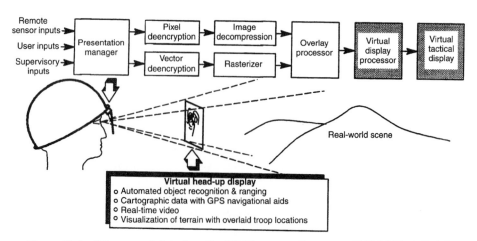

Figure 10.8   Diagram of virtual reality HUD system. Courtesy of NASA Tech Briefs.

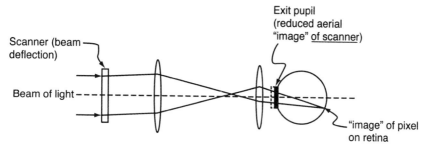

Figure 10.9 Ray tracing of simplified version of virtual retinal display. After Kollin [3], by permission of Human Interface Technology Lab., Wright Field.

wide FOV cholesteric LC collimator. The image source is a 1-in. monochromatic CRT with a fiberoptics faceplate, and color is achieved by placing a switched RGB color filter in front of the white phosphor CRT faceplate. The switching is achieved by having two surface-stabilized ferroelectric LC (SSFLC) shutters operating at 180 Hz, in synchrony with the basic 60-Hz rate. Further developments are in progress for a full-peripheral-vision, high-resolution, full-color HMD using this basic approach and leading to a unit with a FOV of 220° H × 120° V with 84° stereo overlap, integrated with a 1280 × 1024 AMLCD FPDs with a color shutter system. It should be noted that these units are what is termed "fully immersible" or "opaque" units in that the user sees only what is on the display, and the real world is represented only by simulated images as shown in Figure 10.7. The other approach, where the images are in stereo and combine with the real world, is termed "transparent" or "see-through" because the images created by the VR system appear to be superimposed on the real world. This is a popular

TABLE 10.3   Military HMD Display Parameters

Parameter	Unit 1	Unit 2
Field of view (FOV)	40° circular (100% overlap)	60° circular (100% overlap)
	60° H × 40° V (20% overlap)	100° H × 60° V (20% overlap)
Exit pupil diameter (mm)	15	15
Eye relief distance (mm)	> 25 (eyeglass-compatible)	> 25 (eyeglass-compatible)
Transmission	See-through > 24%	See-through > 10%
Luminance (FL)	> 6 (at eye)	> 6 (at eye)
Resolution (arcmin)	2.7 (0.0008 in. linewidth)	4 (0.0008 in. linewidth)
Contrast ratio	> 20	> 20
Video formats	1280 × 1024 at 2 : 1 interlaced	Same
	1024 × 1024 at 2 : 1 interlaced	Same
	640 × 480 noninterlaced	Same

Source: Kaiser Electro-Optics, SIM EYE HMD system.

approach for a number of the VR applications such as automotive panels, some forms of medical imaging, entertainment, and games. The displays used for these types of VR applications are covered later in this section. At this point the discussion of the fully immersible unit is continued.

This type of unit is also available for personal use in a similar unit that differs from the one described previously in that the display consists of a full-color AMLCD with 464,600 color elements in 154,866 color groups. Units similar to it and the other two examples are also available from other suppliers, and the performance capabilities represented by the parameters shown in Table 10.3 may be considered as representative. One example, intended for the entertainment market, has the parameter values shown in Table 10.4. It is clear that the requirements are much less demanding for this type of fully immersible unit than for the militarily oriented application.

The transparent type, on the other hand, uses glasses in which the lenses are coded by color or polarizing filters to respond only to the view that corresponds to the proper eye. The VR scene is then generated in two forms by switching, as for the immersible system; the main difference between the two systems is that the glasses do not entirely block any other view in the transparent systems, and the 3D imagery can be superimposed on the real world. One example of an application for this type of transparent system is an automotive 3D HUD in which the HUD imagery appears on the windshield, and the real world can be viewed through the windshield. Depending on the complexity of the 3D imagery, the resolution can range from a standard 620×480 to the maximum that the projection system can achieve. In this respect, the automotive system is subject to some of the same limitations as the airborne projection HUD, but not the ones that use LCD panels. A tentative set of parameter values is shown in Table 10.5. This is the simplest version of such a driver's aid, and it may be considerably complicated by adding a head tracker so that the 3D imagery can change to correspond to the direction in which the driver is viewing the real world. It is

TABLE 10.4   Entertainment HMD Display Parameters

Parameter	Value
FOV	84°
Display type	TFT LCD
Matrix (V $\times$ H)	720 $\times$ 240
Dot pitch (V $\times$ H, mm)	0.365 $\times$ 0.158
Color pixel format	RGB delta
Response time (ms)	80
Contrast ratio	30
Display dots	172,800
Video signal	Composite analog RGB

Source: Liquid Image Corporation.

TABLE 10.5   Automotive 3D HUD Display Parameters

Parameter	Value
Luminance (fL)	10–20
Resolution (H × V)	640 × 480
FOV°	80–120
Colors (no.)	16–256

also possible to conceive of a totally synthetic system in which the actual scene is artificially generated at night or in low-visibility conditions.

Finally, there is the wide FOV, full-color, high-resolution unit, that is part of the complete airborne system shown in block diagram form in Figure 10.10. The total system is shown, with the HMD restricted to the cockpit, which contains the sensor for the head tracker and the complete eye tracker. This diagram is an extension of the one shown in Figure 10.3, and includes the tracker circuitry, discussed in detail in the next section. The fields of view for the left and right eyes are shown in Figure 10.11, and the resultant requirement for the binocular high-resolution FOV is in the order of 5 arcmin per pixel in the background field, and 1 arcmin per pixel in the high-resolution field. These are rigorous requirements and necessitate the use of a high-resolution, three-CRT projector for each eye, as shown in Figure 10.11. This design is an example of the performance needs of HMD systems. The parameter values are in Table 10.6.

It should be apparent from the variety of HMDs and other means for achieving the display capability needed to implement VR that this is a rapidly advancing technology and will lead to closer and closer approximation of the real world. The ultimate result of this effort, coupled with the concomitant activity in improved software and computer performance, should lead to the

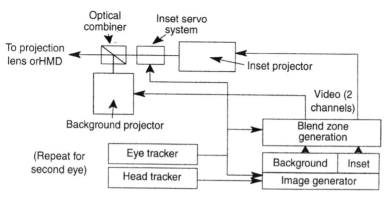

Figure 10.10   Airborne helmet mounted display system. After Fernie [4], by permission of SID.

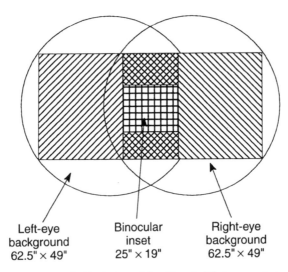

Left-eye
background
62.5" × 49"

Binocular
inset
25" × 19"

Right-eye
background
62.5" × 49"

Figure 10.11   System field of view. After Fernie [5], by permission of SID.

dreamworld of the animation artist, where the real world is replaced for many applications by the fantasy world created by the artists and engineers involved. Whether this is good for civilization remains to be determined, but as a minimum, it should allow great improvements in the design and utilization of most of the applications discussed previously as well as many not even considered at present. These futuristic expectations are part of what is expected in the world and may be considered as promising significant improvements in living conditions.

10.3.2.3   Trackers

As can be seen from Figure 10.11, there may be as many as two trackers needed in a fully instrumented HMD system, that is, one for the eyes and

TABLE 10.6   System Specification for Dual-Resolution HMD

Parameter	Value
Background FOV (per eye)	62.5° (H) × 49° (V)
Background FOV (total)	100° (H) × 49° (V)
Inset FOV	25° (H) × 19° (V)
Background resolution	12 arcmin
Inset resolution	6 arcmin
Luminance	20 fL
Contrast ratio	20
Eye relief	42 mm

Source: Fernie [5].

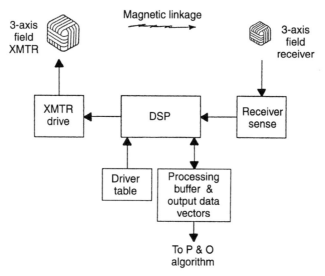

Figure 10.12 Diagram of magnetic head tracker. After Aspin [6], by permission of SID.

one for the head. The head tracker is most commonly found when only one tracker is used. However, the principles involved in all except the movement sensors are quite similar for both types, so a discussion of the head tracker contains much that is relevant to the eye tracker. Therefore, that unit is concentrated on first. Head trackers using ac magnetic technology have been the most successful design because there is no line-of-sight limitation as may occur with optical trackers. One example of the front end of such a unit, using a digital signal processor (DSP) to handle the signals from the magnetic pickups, is shown in block diagram form in Figure 10.12. The magnetic transmitter three-axis coil assembly is energized by the transmitter drive and then transmits the resultant signals by magnetic linkage to the receiver assembly located on the head being tracked, which in turn sends the signals to the DSP, where the position and orientation (P & O) of the head are calculated by means of the P & O algorithm. Representative operating characteristics for this type are given in Table 10.7.

Although this description of an HMD head tracker applies basically to an airborne unit, the same principles may be applied to other types of VR systems that use head movement to determine what the orientation of a VR image should be. Therefore, it may be considered at this point that all relevant aspects of the different types of VR systems that apply to the display application have been covered, at least to a sufficient extent to illustrate what the performance capabilities are and how the systems may be used. Therefore, any further discussion of the nondisplay aspects of the VR application is redundant, and this section concludes the material on VR applications.

TABLE 10.7 Operating Characteristics for VR Head Tracker

Parameter	Value
Update rate (Hz)	120
Latency (ms)	4
Accuracy (< 30-in. range)	0.03 in. RMS,[a] 15° RMS
Resolution	0.0002 in./in. range; 0.05°
Range (ft)	10

[a]Root mean square.

Source: Aspin [6].

REFERENCES

1. Baum, D. R., "Virtual Reality: How Close Are We?" *SID 1993 Dig.* **1**, 754.
2. King, P., "Integration of Helmet-Mounted Displays into Tactical Aircraft," *SID 1995 Dig.* **1**, 663.
3. Kollin, J. S., "The Virtual Retinal Display," *Human Interface Technol. Lab.* **1**, 3, 1992.
4. Fernie, A., "Helmet-Mounted Display with Dual Resolutions," *SID 1995 Appl. Dig.* **1**, 37.
5. Fernie, A., "Helmet-Mounted Display with Dual Resolutions," *SID 1995 Appl. Dig.* **3**, 40.
6. Aspin, W. M., "Crusader Integrated Day/Night Helmet Vision System for Tactical Ejection-Seat Aircraft," *SID 1995 Dig.*, 656.

11

SUMMARY

11.1 INTRODUCTION

As a final review of display applications versus relevant display requirements, it is of interest to examine the data given in Table 11.1. This table represents a compilation of the parameter information given in the previous chapters. It demonstrates in tabular form the relationship between electronic display applications and requirements, as defined by evaluation numbers that range from 1 to 6, with the key for each parameter given at the bottom of the table. Thus, the actual parameter range defined by each number is shown so that the numerical value may be determined for each parameter. Further, the table presents an evaluation summary of the data given in previous chapters for the significant applications covered in this volume, and may be used to arrive at performance ranges for each parameter in relation to each display application.

The parameters of prime significance for each application are listed in Table 11.2.

The evaluation numbers are the same as for Table 11.1, but are shown only for the parameters that are of prime significance for the specific applications. The others that contain the x designation are not intended to be ignored or neglected, but are of somewhat lesser importance in terms of their impact on the operational characteristics of the electronic display portion. That is to say, those parameters with the x designation can be allowed to deteriorate from the values given in Table 11.1 without undue effect on the total system performance from the user's point of view. These data are presented as a guide to acceptable performance levels for the electronic displays for each application listed. Of course, if the values presented in Table 11.1 can be maintained for all applications, the results should be optimum, but if some reduction is necessary, then the x group should be changed before those with numerical values. However, it should be recognized that they are not hard and fast, and it is ultimately up to the user to

TABLE 11.1 Display Applications versus Parameter Requirements

Application	Color	CR	Lum.	Res.	Size	Speed	Gray	Matrix	Element
Avionics	3	3	4	3	4	4	5	5	5
CAD/CAE	4	2	3	3	4	3	2	3	3
CAM	3	2	2	2	3	3	2	3	3
Control	2	2	2	1	6	2	2	2	2
Desktop publishing	5	4	4	5	4	4	3	4	4
Education	2	3	2	2	2	2	3	2	2
Electronic games	5	4	3	3	3	3	3	5	4
Fine art	6	4	4	6	6	3	5	5	5
Geographic (GIS)	3	2	2	3	5	2	4	3	2
Imaging	5	4	4	6	6	3	3	5	5
Graphics art	5	3	4	3	4	2	4	5	4
Instruments	2	4	3	4	3	2	3	3	2
Measurements	3	3	3	4	4	3	3	4	3
Medical	4	2	2	2	4	2	3	2	2
Military	3	5	4	3	5	4	3	5	4
Monitoring	3	4	4	3	5	3	3	3	2
Multimedia	4	2	3	2	4	3	4	2	2
Navigation	3	4	4	3	3	4	3	3	2
Presentations	3	2	3	3	6	3	3	3	2
Simulation	3	3	4	3	6	4	3	3	2
Television	5	3	3	3	6	3	3	3	2
Transportation	3	2	3	2	6	2	2	3	2
Virtual reality	4	3	4	4	1	3	2	3	2
Visualization	5	2	3	4	4	3	5	3	2

Parameter	Key
Colors (no.)	0 = none, 1 = 2–4 bits, 2 = 4–8 bits, 3 = 8–16 bits, 4 = 16–24 bits, 5 = > 24 bits
Contrast ratio (CR)	1 = low (< 5), 2 = med. (5–20), 3 = high (> 20)
Luminance (Lum.) (L = nts)	1 = low (< 20), 2 = med. (20–100), 3 = med. high (100–1000), 4 = high (> 1000)
Resolution (Res.) (TV lines)	1 = low (< 256), 2 = med. (256–512), 3 = med. high (512–1024), 4 = high (> 1024)
Screen size (diagonal in.)	1 = < 5, 2 = 5–12, 3 = 12–16, 4 = 16–21, 5 = 21–27, 6 = > 27
Speed (μs)	1 = > 100, 2 = 10–100, 3 = 1–10, 4 = < 1
Gray scale (Gray) (bits)	0 = none, 1 = 2–4, 2 = 4–8, 3 = 8–16, 4 = 16–24, 5 = > 24
Matrix (W × H)	1 = 620 × 480, 2 = 800 × 600, 3 = 1024 × 756, 4 = 1024 × 1024, 5 = > 1024 × 1024
Element size (mm)	1 = 0.02–0.04, 2 = 0.04–0.08, 3 = 0.08 = 0.16, 4 = 0.16–0.2, 5 = 0.2–0.4

TABLE 11.2   Requirements

Application	Color	CR	Lum.	Res.	Size	Speed	Gray	Matrix	Element
Avionics	x	x	x	x	4	4	5	5	5
CAD/CAE	4	x	x	x	4	x	x	x	x
CAM	3	x	x	x	3	3	x	3	3
Control	x	x	x	x	6	x	x	x	x
Desktop publishing	5	4	4	5	4	4	x	4	4
Education	x	3	x	x	x	x	3	x	x
Electronic games	5	4	x	x	x	x	x	5	4
Fine art	6	x	x	6	6	x	5	5	5
Geographic (GIS)	x	x	x	x	5	x	4	x	x
Graphics art	5	x	4	x	4	x	4	5	4
Imaging	5	x	5	x	4	x	3	5	5
Instruments	x	3	4	x	x	x	x	x	x
Measurements	x	x	x	4	4	x	x	4	x
Medical	4	x	x	x	4	x	3	x	x
Military	x	5	4	5	5	4	x	5	x
Monitoring	x	4	4	x	5	x	x	x	x
Multimedia	4	x	x	x	4	x	4	x	x
Navigation	x	4	4	x	x	4	x	x	x
Presentations	x	x	x	x	6	x	x	x	x
Simulation	x	x	4	x	6	4	x	x	x
Television	5	x	x	x	6	x	x	x	x
Transportation	x	x	x	x	6	x	x	x	x
Virtual reality	4	x	4	4	x	x	x	x	x
Visualization	5	x	x	4	4	x	5	x	x

determine what is most appropriate for the application that is under consideration. With these caveats, the information presented here should be useful in establishing the proper requirements.

11.2   APPLICATIONS VERSUS DISPLAY TECHNOLOGIES

A number of display technologies are discussed in Chapter 2, and it is the purpose of this section of this chapter to review the characteristics of the technologies in terms of their value to the specific electronic display applications in order to establish which characteristics are most important, and what technologies provide the best performance as required to meet the operational technology requirements. As a preliminary to the discussion of the individual electronic display applications, it is useful to repeat the list of applications found in Tables 11.1 and 11.2, in combination with the preferred display technology. However, in order to simplify the analysis, it is useful to repeat the lists of display technologies found in Table 2.1 in slightly modified form at this point.

The major display technologies are listed in Table 11.3a. The other technologies that have been tried, some with only limited success, and others with promising results, are listed in Table 11.3b. It should be recognized that at least some of these technologies are still being investigated or are under advanced development, so there may be significant improvements in the performance of display products using these technologies. Therefore, they should not be overlooked if they exhibit characteristics that may be particularly advantageous for some of the applications. These technologies are not used for any significant applications, and further discussion of their characteristics does not appear to be warranted.

The display applications are shown in Table 11.4, along with the preferred technologies.

It should be noted that several of the technologies listed in Table 11.3b do not appear in Table 11.4; these technologies either (1) have proved unsuccessful despite considerable effort or (2) are too new to have fully demonstrated their capabilities. Examples of group 1 are ECH, EP, and the laser, whereas FET is a prime example of group 2. Of these—although there have been cases of unexpected recovery—it appears unlikely that technologies that have been investigated for a considerable period of time will become successful, although lasers show some signs of comeback. In any event, there are a sufficient number of technologies to meet any need and a variety of formats, so that none of the applications should lack at least one preferred technology for the electronic displays it requires.

TABLE 11.3   Electronics Display Technologies

Technology	Type	Drive
a. Major Technologies		
Cathode ray tube (CRT)	Light–emitting	Ac, dc
Electroluminescent (EL)	Light–emitting	Ac, dc
Light-emitting diodes (LED)	Light–emitting	Dc pulse
Liquid crystal (LC)	Passive	Ac pulse
Plasma (PL)	Light–emitting	Ac, dc
Vacuum fluorescent (VF)	Light–emitting	Ac pulse
b. Other Technologies		
Cathode ray tube (FPC)	Light–emitting	Ac pulse
Electrochemical (EC)	Reflective	Dc pulse
Electrochromic (ECH)	Passive	Dc
Electrophoretic (EP)	Passive	Dc
Electromagnetic (EM)	Reflective	Dc pulse
Field emission (FE)	Light–emitting	Ac pulse
Incandescent (IN)	Light–emitting	Ac
Laser (L)	Light–emitting	Ac pulse
Particle (P)	Reflective	Dc

TABLE 11.4   Application versus Technologies

Application	Technologies	Advantages
Avionics	CRT, IN, LC, VF	High luminance, small space
CAD/CAE	CRT, EL, LC, PL, VF	High luminance, high resolution, FPD
CAM	CRT, LC, VF	High luminance, FPD
Control	CRT, IN, LC, VF	High luminance, small space
Desktop publishing	CRT, LC	High luminance, high resolution, FPD
Education	CRT, LC	High luminance, FPD
Electronic games	CRT, LC	High luminance, FPD
Fine art	CRT	High luminance, high resolution
Geographic (GIS)	CRT, EL, LC, PL	High luminance, high resolution, FPD
Graphics art	CRT, LC	High luminance, high resolution, FPD
Imaging	CRT, EL, LC, PL	High luminance, high resolution, FPD
Instruments	CRT, LC, VF	High luminance, high resolution, FPD
Measurements	IN, LC, LED, VF	Small size, AN
Medical	CRT	High luminance, high resolution
Military	CRT, IN, LC, LED	Size range, long life, high reliability
Monitoring	CRT, IN, LC, LED	Size range, high reliability
Multimedia	CRT, LC	High luminance, high resolution, FPD
Navigation	CRT, IN, LC, LED	Size range, high reliability
Presentations	CRT, LC	High luminance, high resolution, projection
Simulation	CRT, LC	High luminance, high resolution, projection
Television	CRT, LC, PL	Size range, high luminance, projection
Transportation	CRT, IN, EM	Large size, high luminance
Virtual reality	CRT	Stereo, high luminance, high resolution
Visualization	CRT, LC	High luminance, high resolution

11.3   TECHNOLOGY REVIEW

The major technologies that are in use for electronic displays are the ones listed in Table 11.3a, and this review is essentially restricted to these six, with the addition of FET as a promising new technology. Detailed information on these, as well as the others listed in Table 11.3b, may be found in Chapter 2, and the discussion presented here is intended only to highlight the characteristics of each of the most successful technologies in terms of the advantages of the technologies in terms of the application requirements. This is done to a limited extent in Table 11.4, and the following material expands somewhat on the material contained there. To this end, each of the main technologies is listed in Table 11.5 with special attention to rating the primary characteristics in terms of the general display requirements.

It is apparent from these data that the CRT technology is still best for essentially all of the parameters shown in Table 11.5, with the FPD technologies fairly equal in their parameter values. This is because the essential differences do not show up until the products that use these technologies are

TABLE 11.5   Display Technologies Characteristics

	Characteristics					
Technology	Lum. (nits)	CR	Res. (TV)	Response (ms)	Colors (No.)	Pixel Size (in.)
CRT	25–6000	10–50	500–1600	0.1–1	16–16 million	0.025–0.039
LED	30–300	5–25	200–500	0.1–0.5	5–10	0.1–1
EL (ac)	25–75	15–20	500–1500	1.0–10	16–256	0.01–0.05
EL (dc)	50–100	15–20	320–640	1.0–10	1–16	0.01–0.05
Plasma (ac)	15–150	10–20	500–1500	1.0–10	16–256	0.01–0.1
Plasma (dc)	50–250	10–150	500–1000	1.0–10	16–256	0.01–0.1
VFD	50–500	10–50	200–600	0.1–1	16–256	0.20–0.6
LCD	NA	10–20	500–1500	50–500	16–16 million	0.01–0.05

evaluated, which is done in the next section. This may lead to potentially misleading conclusions based on the similarities and the limitations of the data should be borne in mind before jumping to possibly incorrect conclusions about which technologies are best for specific applications. However, the data do indicate that all of the FPD technologies have good potential insofar as their actual parameter values are concerned, and it is the capabilities and limitations that exist in the products that have the major impact on the technology acceptance. However, it is still of some interest to compare the display applications with the required parameter ranges, and this is done in Table 11.6, which summarizes the data presented in the designated chapters.

Many of the applications require the same or very similar parameter values, so the same electronic display equipment and systems may be quite adequate for a large number of the total group of electronic display applications. This should make it possible for the vendors to offer the same products to many of the user groups, which should make it feasible to keep costs down to a minimum. This list should be useful as a concise summary of the parameter requirements for the bulk of the electronic display applications covered in this volume. However, it should be noted that it is offered only as a starting point for the specification and selection of the appropriate hardware, and it is necessary to examine the exact operational requirements for each specific application in order to arrive at a truly optimum choice. In addition, other parameters, such as cost, reliability, power consumption, and availability, are also of importance, but apply to the total equipment and system, and are not included here for the displays alone.

11.4   PRODUCT REVIEW

The main electronic display products of interest are those described in Chapters 3 and 4 and listed in Table 11.7 for dynamic types and Table 11.8 for hard-copy types.

TABLE 11.6   Application Parameter Requirements

Application	Lum. (nits)	CR	Res. (TV)	Speed (in./s)	Colors (No).	Pixel Size (in.)
Avionics	20–6000	5–20	525–1280	1–10	1–256	0.002–0.1
CAD/CAE	100–300	10–30	480–1024	1–10	256–16 million	0.001–0.01
CAM	25–50	5–15	480–1024	1–20	16–256	0.001–0.01
Control/monitoring	25–50	5–15	480–1024	5–10	4–16	0.005–0.01
Desktop publishing	25–50	5–10	480–1024	2–20	256–16 million	0.001–0.01
Education	25–50	5–10	480–1280	1–10	256–16 million	0.001–0.01
Electronic games	30–300	10–30	480–1280	1–20	16–256	0.001–0.01
Fine art	30–300	10–50	1024–3600	1–10	256–16 million	0.001–0.01
GIS	35–70	10–20	480–1280	1–10	1–16	0.001–0.01
Graphics art	30–300	10–50	480–1600	1–20	256–16 million	0.001–0.01
Imaging	100–300	10–30	600–1280	1–10	256–16 million	0.001–0.01
Instruments	30–100	10–20	500–1000	10–20	1–16	0.05–0.5
Measurements	30–100	10–20	480–1024	10–20	1–16	0.01–0.05
Medical	100–300	10–20	800–1600	1–10	16–16 million	0.001–0.01
Military	100–3000	20–50	480–1280	1–10	4–256	0.001–0.01
Multimedia	50–300	10–50	600–1280	1–10	256–16 million	0.001–0.01
Navigation	35–100	10–20	620–1280	1–10	4–16	0.001–0.01
Presentations	35–1000	10–30	640–2048	1–20	16–16 million	0.001–0.01
Simulation	40–300	10–20	480–1600	10–20	16–16 million	0.001–0.01
Telecommunication	25–50	10–20	480–1024	5–10	16–256	0.01–0.1
Transportation	50–100	10–30	480–1024	1–10	16–1024	0.001–0.1
Utilities	25–100	10–30	480–1024	1–20	16–1024	0.001–0.1
Virtual reality	20–100	10–30	480–1280	1–10	256–16 million	0.001–0.01
Visualization	50–100	10–20	600–1280	10–20	256–16 million	0.001–0.01

TABLE 11.7   Dynamic Electronic Display Products

Product	Lum. (nits)	CR	Res. (TV)	Speed (ms)	Colors (No.)	Pixel Size (in.)
CRT monitor	30–3000	10–50	480–1600	0.01–1	256–16 million	0.022–0.039
EL monitor	30–100	5–15	480–1024	0.1–1	16–256	0.06–0.2
Ac EL panel	30–100	10–20	480–1024	0.1–1	16–256	0.02–0.2
Dc EL panel	30–300	5–15	480–620	0.1–1	16–256	0.01–0.1
Plasma monitor	25–50	5–15	480–1024	0.2–1	16–256	0.02–0.2
Plasma panel	25–50	5–20	480–1024	0.2–1	16–256	0.02–0.2
VF panel	150–1000	30–50	480–620	0.01–0.05	16–256	0.02–0.05
LCD monitor	NA	10–20	480–1280	1–10	256–16 million	0.01–0.1
LCD panel	NA	10–20	480–1024	10–100	16–256	0.01–0.1

The parameters for the plotters and/or printer–plotters presented in Table 11.8 cover only the most commonly used types. The parameters that are used are those of main interest; there may be others that are of concern to some users, but not enough to warrant inclusion.

Other parameters that may be included are the accuracy and repeatability for the flatbed, drum, and electrostatic plotters, which ranges from 0.1–0.5%

TABLE 11.8   Hard-Copy Display Products

Product	Resolution	Pens (No.)	Speed
Flatbed	0.0002–0.0035 in.	1–8	10–30 ips
Drum	0.00006–0.005	1–14	3–47 ips
Electrostatic	200–400 dpi	1–4	0.3–2.5 ips
Inkjet	130–360 dpi	12–64 jets	167 cps
Laser	240–600 dpi	1–4 beams	4–120 ppm
Thermal	100–400 dpi	1 printhead	NA
Film-based	800–8000 TV	1–3 beams	30 s–5 min

of the total movement for the accuracy, 0.0004–0.005 in. for the repeatability of the first two, and 0.1–0.2 for the accuracy of the third. As noted previously, other parameters may be of interest for the other types, and may be obtained from the manufacturers, if desired.

11.5   APPLICATIONS VERSUS DISPLAY PRODUCTS

The list of applications versus the preferred display technologies is given in Table 11.9, and it should be recognized that this list is subject to constant change as new products appear and old ones are superseded. Thus, the list must be used circumspectly when deciding on which products are most appropriate for each application. However, it may be used as a starting points, with additions and corrections added as they become available.

The list of advantages is limited to those of most significance, but can be used as a guide to what may be achieved with each selected product. In addition, the actual products may be expanded to include all the relevant technologies shown in Table 11.5 by comparing the data in that table with the desirable advantages listed in Table 11.9.

11.6   EVALUATION RATING MATRICES

The final summary data for this volume are included in the following matrix tables that rate on a numerical basis of 1–10 all applications listed in the tables found in this chapter, in terms of the technologies and products covered in tables in previous chapters. The numerical evaluation is intended to supplement the selection choices recommended in Tables 11.4 and 11.9, and establish a somewhat more exact basis for arriving at optimum choices for the technologies and products best suited to the specific applications. Tables 11.10 and 11.11, the final tables in this chapter, complete the review and summarization of the material found in more detail in the previous chapters. CRT technology scores highest for all applications because, with

TABLE 11.9   Applications versus Display Products

Application	Products	Advantages
Avionics	Monitors	High resolution, color
CAD/CAE	Monitors, panels	High resolution, color
CAM	Panels	Medium resolution, size
Control	Panels	Medium size
Desktop publishing	Monitors, panels	High resolution, color
Education	Panels	Medium resolution, color, size
Electronic games	Panels	Color, size
Fine art	Monitors	High resolution, color, size
Geographic (GIS)	Monitors, panels	High resolution, color
Graphics art	Monitors	High resolution, color, size
Imaging	Monitors	High resolution, color
Instruments	Panels	Medium resolution, size
Measurements	Panels	Medium resolution, size
Medical	Monitors, panels	High resolution, color
Military	Monitors, panels	High resolution, color, size
Monitoring	Panels	Medium resolution, size
Multimedia	Monitors, panels	High resolution, color, size
Navigation	Monitors, panels	Medium resolution, color
Presentations	Monitors, panels	High resolution, color, size
Simulation	Monitors	High resolution, color, size
Television	Monitors, panels	High resolution, color, size
Transportation	Panels	Medium resolution, size
Virtual reality	Monitors	High resolution, color, size
Visualization	Monitors	High resolution, color, size

the exception of display size, power, and portability, it provides the best performance for all parameters listed in Table 11.7. In addition, the CRT cost is significantly lower for each application, as is more apparent in the product evaluations shown in Table 11.11.

Again, the CRT technology as used for the CRT monitor achieves the best total score, with LCD monitors not too far behind, and EL and PL monitors still reasonable choices. However, it should be recognized that only avionics, measurements, and some aspects of the military provide a premium for size and portability among the listed applications, although many of the others might benefit to some extent by the use of portable and flat-panel electronic displays, assuming that the higher price of non-CRT units is not a deterrent. However, in general, it can be assumed that special considerations are needed to justify the use of the non CRT units, such as the need to work on airplanes or elsewhere in the field, which depend on situations that are unique to the user rather than general in terms of the applications. With these and similar considerations in mind, it should be practical to use the data found in Table 11.10 as a starting point in the selection of the best, or at

TABLE 11.10 Numerical Rating Matrix: Applications versus Technologies

Application	Technology							
	CRT	Ac EL	Dc EL	LED	PL	VF	AMLCD	PMLCD
Avionics	10	3	4	6	4	5	8	7
CAD/CAE	10	4	4	2	5	3	9	7
CAM	8	4	4	4	3	4	8	6
Control	7	6	6	5	4	3	7	5
Desktop publishing	10	7	7	2	8	4	9	7
Education	9	6	6	1	7	3	8	6
Electronic games	8	5	5	2	5	3	9	8
Fine art	10	8	7	0	7	4	9	8
Geographic (GIS)	10	6	6	1	6	5	8	6
Graphics art	10	5	4	0	5	3	8	6
Imaging	10	8	7	2	8	4	8	6
Instruments	9	7	7	3	7	5	9	8
Measurements	8	6	6	4	6	6	8	7
Medical	8	6	6	2	7	3	7	6
Military	10	8	9	6	9	7	10	8
Monitoring	9	7	6	4	8	6	8	7
Multimedia	10	7	7	2	8	7	9	7
Navigation	8	6	5	3	7	5	8	6
Presentations	10	7	7	4	8	4	9	7
Simulation	10	7	6	2	7	3	9	7
Television	10	8	7	1	9	4	10	8
Transportation	8	6	5	3	7	4	7	6
Virtual reality	10	7	6	2	8	3	8	6
Visualization	10	7	5	1	8	3	9	7

least the adequate, choice for the electronic displays needed for each application listed.

11.7 CONCLUSIONS

The following conclusions may be derived from the information contained in the previous chapters of this volume:

1. There are at least eight fully developed technologies that are in extensive use for the design and manufacture of electronic displays. These are listed in Table 11.10.
2. Similarly, there are eight fully developed products available and in fairly extensive use for at least one, and generally more than two, of the applications. These are listed in Table 11.11.
3. Among the other technologies that have been investigated and developed to varying degrees of completion, and with varying degrees of real

TABLE 11.11   Numerical Rating Matrix: Applications versus Products

Application	CRT Monitor	EL Monitor	EL Panel	PL Monitor	PL Panel	VF Panel	LCD Monitor	LCD Panel
Avionics	10	3	4	6	4	5	8	7
CAD/CAE	10	4	4	2	5	3	9	7
CAM	8	4	3	4	3	3	8	6
Control	7	5	4	5	4	3	7	5
Desktop publishing	10	6	5	7	7	3	9	7
Education	8	5	5	8	7	2	8	6
Electronic games	8	4	3	5	5	3	9	8
Fine art	10	6	6	7	7	3	8	8
Geographic	9	6	6	8	6	4	8	6
Graphics art	10	5	4	8	7	3	8	6
Imaging	10	7	7	8	8	3	8	6
Instruments	7	6	6	7	7	4	8	8
Measurements	7	5	5	5	5	4	7	7
Medical	8	5	5	7	6	4	7	6
Military	10	7	6	6	7	6	10	8
Monitoring	9	7	6	4	8	6	8	7
Multimedia	10	7	7	2	8	7	9	7
Navigation	8	6	5	3	7	5	8	6
Presentations	10	7	7	4	8	4	9	7
Simulation	10	7	6	2	7	3	9	7
Television	10	7	7	8	7	4	10	8
Transportation	8	5	5	6	6	4	7	6
Virtual reality	10	6	6	6	5	3	8	6
Visualization	10	6	5	6	5	3	8	7

or potential success, at present only the field emission, and possibly the laser approaches, show meaningful promise of ultimate success. These are described and discussed to varying levels of detail in Section 2.3.3.9 of this volume.

4. The use of hard-copy devices continues to a surprising extent despite the predicted trends to a paperless society.

There does not seem to be any strong push to eliminate hard-copy devices (item 4, above) to any significant extent in the near future, but the advances in storage techniques such as the development of high-capacity CD-ROMs and rewritable high-density CDs and DVDs should make it feasible to minimize the need for archival storage and the distribution of hard-copy material, so these devices may become unnecessary. However, this prediction has been made too often without verification to be reliable, and firm statements as to the future of hard copy should be avoided. To this end, the available technologies and products of this type are discussed in detail in Section 3.4 of this volume, although no attempt has been made to produce an evaluation matrix. This is because hard copy is usually less critical to the success of the display aspects of the applications, and the choice of technology and product type is frequently more affected by cost and aesthetics than by considerations of engineering, science, and operational need. Therefore, any attempt to provide evaluation matrices in terms of applications has been avoided and left as an exercise for the user.

With these comments and conclusions, this volume on applications for electronic displays concludes, with the hope and expectation that it is a useful addition to the rather small group of texts concerned with the general topic of electronic displays.

INDEX

Entries preceded by (T) indicate tables.

Advertising, 6
Animation, 7
Application display specification, art, 19–21, (T) 20
Application requirements, 6
Art, 7
Avionics, 7

Bar-graph parameters, (T) 97
Business systems, 8. *See also* Computer aided design, engineering, and manufacturing (CAD, CAE, CAM)

Candela, 26
CIM, 9
Colloidal display, 69
Color CRTs, 41–46
Computer-aided design/manufacturing (CAD/CAM), 8
Computer-aided design, engineering, and manufacturing, 289–312, (T) 301
  application descriptions, 290–301
  block diagrams, 307–311
  CAD/CAE, 302–304
  CAM, 304–307
  CAM outputs, (T) 300
  computer-aided design and engineering (CAD/CAE), 292–295
  computer-aided manufacturing, 296–301

graphics-workstation specification, (T) 295
hard copy display specification for CAD/CAE applications, (T) 306
processor and dynamic display specification CAD/CAE application, (T) 302
  CAM application, (T) 307
Computer graphics, 207–228, (T) 208
  advertising, 208–209
  animation, 209–210
  applications, (T) 208
  application descriptions, 208–215
  block diagrams, 225–228
  computer-based. 226
  desktop publishing systems, (T) 211
  fine art, 212–213
  geographic information systems, 213
  other applications, 226
  simulation, 213–214
  sports, 214
  television, 214–215
Computer-graphics communications, control, 9
Computer-integrated manufacturing, *see* CIM
Cones, 25
Contrast, contrast ratio, 27
CRT, 33–46
  beam-forming region, 35
  beam-shaping region, 35–37
  beam-deflection region, 37–39
  general description, 33–46
  light-production region, 39–41

CRT monitors, 125–132
  applications, (T) 125
  graphic boards, (T) 131
  graphics color monitor parameters, (T) 131
  raster TV, 127–123
  vector types, 125–127

Data processing, desktop publishing, 10
Display applications
  descriptions of, 4–16
  list of, (T) 4
Display parameters, 17
Document imaging user industries, (T) 243
Document preparation, education, 10

Electrochromic, 66, 69
Electroluminescence (EL), 88, 102–105
Electroluminescence and plasma, 88, 91–96
Electromagnetic, 70–72
Electrophoretic, 69
Electronic games, entertainment, geographic
  systems, 11
Entertainment, 274–288
  application descriptions, 275–288
  arcade games, 276–277
  block diagrams, 283–288
  computer-driven games, 276
  dynamic display specification
    arcade games application, (T) 281
    electronic games application, (T) 281
    handheld electronic games application,
      (T) 280
    PC-operated games application, (T) 281
    video application, (T) 282
  electronic games, 275–278
  handheld games, 275–276
  HMD display parameters, (T) 282
  specifications, 279–282
  video, 278–279, 282
  video sources, (T) 279

Field emission, 65–66
Flat-panel display technologies, (T) 23
Flicker, 31–32
Fovea, 25
Flat-panel display (FPD) monitors, 132–140
  AMLCD color monitor parameters,
    (T) 140
  electroluminescence, 132–134
  EL monitor parameters, (T) 133
  LCD monitor parameters, (T) 139
  liquid crystal (LC), 135–140
  plasma, 134–135
  plasma monochromatic monitor parameters,
    (T) 137

Flat-panel display (FPD) technologies, 46–74,
  (T) 101
  comparison of, (T) 73
  electroluminescence, 50–53
  electroluminescent technologies, (T) 50
  general description, 46–47
  light-emitting diodes, 47–50
  performance capabilities of, (T) 74
  visible LEDs, (T) 49

Gas discharge (plasma), 53–57
  characteristics of gas discharge, (T) 56
Generalized display system, 20

Hard copy, 140–151
  devices, (T) 41
  dot-matrix printer-plotter, 149–151
  electrostatic plotter, 143
  film-based technology, 151
  inkjet printer-plotter, 143–147
  laser printer-plotter, 147–148
  pen plotter, 141–143
  plotter parameters, (T) 143
  technology, 141–151
    parameters, (T) 218
  thermal printer-plotters, 148–149
  units for mainframe computers, (T) 170
Human factors considerations, 23–32
  introduction, 23–25

Illuminance, 27
Imaging systems, 11
Incandescent, 72
Information systems, inspection systems, 12
Input devices, 177–183, (T) 183
  graphics (data) panel, 178
  light pen, 177–178
  mouse, 178–180
  trackball, 180–183

Large-screen systems, 151–161, (T) 152
  CRT projection, 153–155
  front projection, 154–155
  LCD projection panel parameters,
    (T) 161
  light-valve and LCD projection,
    155–161
  liquid crystal, 160–161
  rear mirror, 153–154
  schmidt, 152–153
Legibility, 31
Light emitting diodes (LEDs), 87–90
  large discrete digit display parameters,
    (T) 91
  module parameters, (T) 104

Liquid crystal, 59–65, 97, 99
  applications for readouts, (T) 99
  dot-matrix parameters, (T) 99
Lumen, luminance, 26
Luminous flux, 27

Major technologies, 33–74
Manufacturing, measurement, message
  displays, 12
Military, monitoring, multimedia,
  navigation, 13
Matrix addressing, 76–82
  active, 81–82
  basic techniques, 76–81
Monitors, 125–140
Multimedia and presentations, 255–273
  applications descriptions, 255–258
  block diagrams, 269–273
  briefings, 267–269
  display technologies and products, 258–261
  dynamic display specification
    briefings application, (T) 268
    education application, (T) 263
    kiosk application, (T) 264
    LCD projection system, (T) 267
    medical application, (T) 263
    multimedia and presentations
      applications, (T) 262
    training applications, (T) 265
  dynamic display technology parameters,
    (T) 260
  education, 262–263, 267
  hard copy technologies, 260–261
    parameters, (T) 261
  kiosks, 264–265
  medicine, 263–264
  multimedia, 255–257
  multimedia applications, 262–265, (T) 256
  presentations, 257–258
  presentations applications, 266–269, (T) 258
  presentations requirements, (T) 266
  specifications, 261–269
  technologies, 259–261
  technologies versus products, (T) 259
  training, 265
Multiplexing, 75–76
  and matrix addressing, 74–82

Output devices and systems, 87–164
  applications, technologies, products, (T) 162
  applications versus ratings, (T) 163

Panels, 101–125
  ac plasma, (T) 111
  color STN display parameters, (T) 119

dcel parameters, (T) 109
  electroluminescence, 102, 104–105
  field effect devices (FEDs), 121–122
  field emission display (FED) parameters,
    (T) 122
  light emitting diodes (LEDs), 101–102
  liquid-crystal displays (LCDs), 116–122
  optical parameters, (T) 109
  other technologies, 122–124
  plasma, 105–113
  rapidot display parameters, (T) 124
  refreshed dc dot-matrix plasma panel,
    (T) 111
  vacuum fluorescence, 111, 114–116
  vacuum-fluorescent display (VFD)
    parameters, (T) 116
Photometric, 25–27
  conversion factors, (T) 27
  parameters, 17–19, 26, (T) 17
  terms, 17, 19, 26, (T) 19
  visual parameters, 25–32
Photometry, definition of, 17
Plasma, 88
  bar-graph parameters, (T) 97
  dc, 91–97
  planar, 88, 91
  planar dc display parameters, (T) 94
  raised cathode, 91
  self-scan, 92–97
Presentations, simulation, 14

Readouts, 87–100
  technologies, (T) 88
Resolution, 28–30
  conversion factors for, (T) 29
  modulation transfer function, 26
  shrinking raster, 28
  spot size, 30
  television limiting, 29
Retina, 25
Rods, 25

Simplified display system, 21
Specific applications, list of, (T) 16
Specifications, 217–225
  advertising, 219
  animation, 220
  block diagrams, 217–228
  desktop publishing, 220–221
  dynamic display specification for computer
    graphics
    advertising application, (T) 219
    animation application, (T) 220

Specifications (*Continued*)
desktop publishing application, (T) 221
fine art application, (T) 222
geographic information systems (GIS),
(T) 223
simulation application, (T) 224
sport application, (T) 224
television application, (T) 225
fine art, 221–222
geographic information systems (GIS),
222–223
hard copy display specification for computer
graphics
animation application, (T) 220
fine art application, (T) 222
geographic information systems (GIS)
application, (T) 223
simulation, 223–224
sports, 224–225
television, 225
Sports, 14
Summary, 335–346
application parameter requirements, (T) 341
applications
versus display products, (T) 343
versus display technologies, 337–339
versus technologies, (T) 339
conclusions, 344–346
display applications versus parameter
requirements, (T) 336
display technologies characteristics, (T) 340
dynamic electronic display products, (T) 340
electronics display technologies, (T) 338
evaluation rating matrices, 342–344
hard-copy display products, (T) 342
numerical rating matrix
applications versus products, (T) 345
applications versus technologies, (T) 344
product review, 340–342
product requirements, (T) 337
technology review, 339–340
Suspended particle, 69–70
Systems and user applications, 166–202
ASCII terminal specifications, (T) 173
client-server applications, (T) 175
computers, 169–171
desktop, laptop, notebook, and palm-top
computers, 170–171
mainframes and minis, 169–170
personal computer/workstation versus
display type, (T) 171
terminals, 171–175
workstations, 175–177
workstation specification, (T) 177
X window terminal specifications, (T) 173

Technologies versus products, (T) 215
Technology and product descriptions, 215–217
dynamic display, 216
hard-copy, 216–217
Technology comparisons, 100
advantages and disadvantages, (T) 100
Technology parameters, (T) 217
Television, 14
Transportation, 15

User applications, 207–334
Utilities, 15

Vacuum-fluorescent (VFDs), 57–59, 97–98
A/N display parameters, (T) 99
Virtual reality, 15, 313–334
advertising, 315
applications, 314–321, (T) 314
architecture, 316
automative 3D HUD parameters, (T) 331
CAD/CAM, 316–317
entertainment, 317–318
equipment, 326–334
games, 318
helmet-mounted display (HMD), 327–332
parameters, (T) 330
manufacturing, 319
medical imaging, 319
military HMD display parameters, (T) 329
operating characteristics for VR head
tracker, (T) 334
processor parameters, (T) 327
simulation, 319–320
systems and software, 321–326
system specification for dual-resolution
HMD, (T) 232
technologies, 321–334
television, 320
trackers, 320
Visualization and imaging (VISIM), 229–253
application descriptions, 231–239, (T) 230
CAD/CAE, 232–233
block diagrams, 249–253
display equipment, (T) 231
documents, 233–234
dynamic display specifications
medical application, (T) 247
presentations application, (T) 247
earth resources, 234–236
geographical information systems (GIS),
236, 244
hard-copy display specification for VISIM
slidemaking application, (T) 248

Visualization and imaging (VSIM) (*Continued*)
  imaging, 236, 244–246
    categories, (T) 245
    requirements, (T) 246
  mathematics, 236–237, 247–248
  medical applications, 238, 247
  presentations, 238–239, 247
  scientific data, 239, 248–249
  specifications, 240–249
    CAD/CAE applications, (T) 242
    documents applications, (T) 244
    earth resources application, (T) 244
    geographic information system (GIS)
      application, (T) 245
    mathematics applications, (T) 246
    scientific application, (T) 248
  technologies, 239
  technologies versus products, (T) 232
  user-driven dynamic display technology
    parameters, (T) 240
  user-driven hard copy technology
    parameters, (T) 241

Workstations versus host-terminal systems,
    183–202
  additional display specifications,
    (T) 201
  applications and requirements, (T) 184
  application descriptions, 184–202
  avionics, 184–188
  avionic application
    display specification avionic application,
      (T) 191
    display specification holographic HUD,
      (T) 200
  business graphics, 188
    application, (T) 191
    system requirements, (T) 192
  communication, 190, 192
  desktop publishing and document
    preparation, 192–194, (T) 193
  education, 194, (T) 195
  geographic information systems (GIS),
    (T) 195
  information systems, 194–195
  input devices, (T) 197
  instrumentation, (T) 197
  manufacturing, 198
  manufacturing application, (T) 198
  measurements and instruments, 195–198
  military applications, 199–202, (T) 199
  process control, 201
  simulation, 201–202
  simulation requirement, (T) 202